Electromagnetic Wave Absorbers

Detailed Theories and Applications

Youji Kotsuka

Published by John Wiley & Sons, Inc., Hoboken, New Jersey.
Published simultaneously in Canada.

For general information on our other products and services or for technical support, please contact our Customer Care Department within the United States at (800) 762-2974, outside the United States at (317) 572-3993 or fax (317) 572-4002.

Wiley also publishes its books in a variety of electronic formats. Some content that appears in print may not be available in electronic formats. For more information about Wiley products, visit our web site at www.wiley.com.

Library of Congress Cataloging-in-Publication Data:

Names: Kotsuka, Y. (Youji), 1941- author.
Title: Electromagnetic wave absorbers : Detailed theories and applications / Youji Kotsuka.
Description: Hoboken, New Jersey : John Wiley & Sons, Inc., 2019. | "Published simultaneously in Canada"–Title page verso. | Includes bibliographical references and index. |
Identifiers: LCCN 2019003710 (print) | LCCN 2019011758 (ebook) | ISBN 9781119564140 (Adobe PDF) | ISBN 9781119564386 (ePub) | ISBN 9781119564126 | ISBN 9781119564126q(hardback) | ISBN 1119564123q(hardback) | ISBN 9781119564140q(ePDF) | ISBN 111956414Xq(ePDF) | ISBN 9781119564386q(epub) | ISBN 1119564387q(epub)
Subjects: LCSH: Electromagnetic waves–Transmission. | Absorption.
Classification: LCC QC665.T7 (ebook) | LCC QC665.T7 K68 2019 (print) | DDC 539.2–dc23
LC record available at https://lccn.loc.gov/2019003710

Cover Design: Wiley
Cover Image: © KTSDESIGN/Getty Images

Set in 10/12pt WarnockPro by SPi Global, Chennai, India

Printed in the United States of America.

V10012762_080919

Contents

Preface

The absorption, reflection, and transmission phenomena of electromagnetic (EM) waves are the most fundamental subjects for those involved in EM-wave engineering.

Although this book is entitled "*Electromagnetic Wave Absorber*," it has the nature of an EM-wave theory textbook. And, thus, the following two points have been essentially kept in mind. First, the basic physical phenomena are explained as far as possible before describing the detailed theory in each chapter. Secondly, the derivation processes of each important equation are presented in detail along with the appendix explanations.

The first half of this book contains the foundations of EM-wave engineering, viz, the transmission line theories necessary for EM-wave absorber analysis, the basic knowledge of reflection, transmission, and absorption of EM waves, computer analysis, etc.

Based on this, the second half describes specific mediums, the measurement methods of material constants, absorber application examples, methods of absorber design, autonomously controllable EM-wave absorbers, etc.

Now, what is an EM-wave absorber? First, in Chapter 1, in order to understand the overall picture of EM-wave absorbers, they are classified and arranged in a multifaceted manner, including the history of their development process.

After describing the method of determining the amount of EM-wave absorption, their classifications by constituent material, composition form, and frequency are described.

In order to deliver a comprehensive image, these classifications are presented in a table. In addition, the table also contains some application fields of the EM-wave absorber together with the materials to be used, and a new EM-wave absorber, which is described afterwards. The EM-wave problems can be often treated by replacing them as transmission line problems. By doing so, it becomes easy to understand the phenomena and characteristics in the EM-wave problems. Therefore, in Chapter 2, time is devoted to also deepen the understanding of the transmission line theory, including the derivation process of its relevant equations.

After clarifying the relationship between the reflection coefficient and impedances of the transmission line, the principle of the Smith chart constitution is presented in detail along with an admittance chart. Furthermore, this chapter includes the derivation methods of Maxwell's equations, besides presenting the reflection and transmission phenomena of a plane incident wave for perpendicular and oblique incidence.

Chapter 3 studies the reflection coefficients for the cases of perpendicular and oblique incidence on a flat plate-type single-layer wave absorber and a multilayer wave absorber, respectively. In addition, the theoretical analysis of multiple reflections is introduced in the case of an EM-wave absorber placed in a room.

In recent years, a lot of simulation methods or analyses have been introduced for the analysis of EM-wave absorber characteristics. However, it is important to understand their basic theories first.

Chapter 4 describes two powerful simulation methods, viz, the finite difference time domain (FDTD) and finite element (FE) methods. Here, the theories behind the methods are explained in such a way that the reader can understand them well. Regarding the FDTD method, the evaluation of boundary conditions and the cell size division in an analytical region are shown on the basis of an actual wave absorber analysis data.

Next, since the analysis of the FE method is, in principle, based on the variational method, the concept of the variational method is clearly explained with concrete examples. For the FE method, two approaches have been introduced: (i) variational method using a functional and (ii) weighted residual method, being defined as a direct method, without using the functional. For the latter, the three-dimensional current vector potential method is introduced in detail, and an eddy current absorber is demonstrated as an example.

The characteristic of an EM-wave absorber largely depends on its material. Therefore, two typical EM-wave-absorbing materials, carbon and ferrite, are investigated in Chapter 5 from the viewpoints of their crystal structures.

Recently, material technologies have made remarkable progress and many new materials have been produced. Introduction of these materials based on their implementation concepts is also important for new EM-wave absorber designs. Chapter 6 explains three such media, viz, chiral media, ferrite anisotropic media, and metamaterials as an example of special mediums.

Concerning these subjects, the derivation processes of the theoretical equations along with their physical interpretations are explained, and examples of an EM-wave absorber and attenuator are shown.

To know the measuring method of material characteristics of the EM-wave absorber and absorber characteristics themselves is important in order to grasp in advance the required characteristics being determined. In Chapter 7, which is entitled as Measurement Methods of EM-Wave Absorbers, the measurement methods of the EM-wave absorber material constant and the EM-wave absorber characteristics are introduced. Here, in order to understand the fundamental principles of the measurement method and to enhance applicability, the theories including the fundamental principles of measuring material constants together with conventional methods and the measurement methods of EM-wave absorbers are described in detail.

For future EM-wave absorber development, it is also important to understand what kind of absorber is used in which electromagnetic environment. Chapter 8 describes the materials being used and their composition, focusing on absorbers that have been put into practical use; and detailed data are given, from a general flat plate structure to EM-wave absorbers for use in a building wall.

By the way, the commonly used materials for the EM-wave absorber are a low-conductive material, a carbon material, and a magnetic material (typically, a ferrite), and the like.

When designing a new EM-wave absorber having the desired characteristics using these conventional materials, we face difficulties because the absorber materials have to be made through the process of controlling the mixing ratio of raw materials, firing temperature, pressure, etc. Therefore, in Chapter 9, wave absorbers introducing the new concept of "equivalent transformation method of material constant (ETMMC)" are introduced. Here, one method of constructing a new EM-wave absorber is described, which does not require the complicated steps involved in conventional material design. This chapter describes the method of (i) combining two or three conventional materials divided into macro-sizes using a conventional material, (ii) providing small holes in an EM-wave absorber material, (iii) mounting periodical conductive elements on an absorbing material surface, and (iv) introducing integrated circuit concepts.

Currently, artificial intelligence (AI) technologies have advanced rapidly, and it seems that material technologies should assimilate this trend as one aspect. Therefore, it is necessary to design EM-wave absorbers that can be autonomously controlled electrically. Chapter 10 introduces a new EM-wave absorber that can be controlled electrically, called "autonomous controllable metamaterial (ACMM)" absorber. This is an EM-wave absorber based on a new implementation concept satisfying all conditions to be imposed on an EM-wave absorber, and independent of the oblique incidence and polarization characteristics, etc.

As can be speculated from the descriptions, although this is written as a book on EM-wave absorbers, the detailed explanations provided impart to it the properties of a textbook on EM-wave theories.

In publishing this book, I would like to thank Professor Arye Rosen for his devoted cooperation, valuable advice, and encouragement. I owed Mrs. Daniella Rosen for her heartwarming support during my research activities. I express my sincere thanks for her. I acknowledge my heartfelt gratitude to Professor Andre Vander Vorst for his valuable comments and thoughtful suggestions in this regard.

I am deeply grateful to Professor Kunihiro Suetake and Professor Yasutaka Shimizu of the Tokyo Institute of Technology, who provided the valuable opportunity to study the EM-wave absorber and their kind guidance throughout the duration of the research.

This book contains the research contributions of Dr. Mitsuhiro Amano, and I express my appreciation for his sincere efforts in EM-wave absorber studies. I also express my thanks to Professor Emeritus of Ryuji Koga in Okayama University for his cooperation in this book publication.

The basic structural concept in chapter 10 of this book was underpinned by the cancer treatment research on EM-waves guided by the Founder, President Shigeyoshi Matsumae of Tokai University. I would like to express my sincere gratitude for this valuable guidance.

It will be an unexpected delight for the author if this book would be widely helpful from all students at the university level to researchers who are undertaking EM-wave study fields.

23 January 2019

Youji Kotsuka
Japan

1

Fundamentals of Electromagnetic Wave Absorbers

Needless to say, learning the theory and application of wave absorbers entails learning the fundamentals of electromagnetic (EM)-wave engineering itself. In short, this means learning about a broad range of basic matters such as the following:

(a) Transmission line theory, which will aid in understanding the fundamental phenomena of EM waves;
(b) Analytical methods to learn EM-wave reflection and transmission phenomena;
(c) Various behaviors of EM waves;
(d) Theory of EM-wave analysis by computer simulation;
(e) Basic knowledge of EM-wave materials;
(f) Measurement of EM-wave material constants;
(g) The EM-wave environment associated with wave absorbers;
(h) Fundamental concepts of artificial materials;
(i) Knowledge of EM-wave absorbers that can be assimilated with artificial intelligence (AI) technology, and other matters.

Even if called just an "EM-wave absorber," its application fields are broadly extended. Particularly, in recent years, higher frequency applications in various kinds of communication systems have advanced rapidly. However, as the frequency region becomes higher, measures against EM scattering and diffracted waves are inevitably required.

Also, as is well known, EM waves are widely used in fields ranging from communication technologies to medical applications. Therefore, the existence of a radio wave absorber plays an important role ranging from preservation of such communication environment safety down to human body protection [1].

In this chapter, in order to make it easier to understand the contents of this book, basic matters on EM-wave absorbers are arranged from various perspectives.

Electromagnetic Wave Absorbers: Detailed Theories and Applications, First Edition. Youji Kotsuka.
© 2019 John Wiley & Sons, Inc. Published 2019 by John Wiley & Sons, Inc.

After first defining what an EM-wave absorber is, Section 1.1 briefly describes the history of EM-wave absorber development along with the application fields.

In Section 1.2, the quantitative representation method of the EM-wave absorption characteristic, namely, the reflection coefficient is defined. In Section 1.3, the EM-wave absorbers are classified and described from the viewpoint of appearance, composition form, material, and frequency characteristics; these are summarized in a table. In Section 1.4, various applications of EM-wave absorbers are introduced together with the literature. Finally, new wave absorber technologies described in the later chapters are briefly introduced.

1.1 Introduction to Electromagnetic-Wave Absorbers

As the name of the EM-wave absorber, "radio wave absorber," is often interpreted conventionally. However, the expressions "electromagnetic wave absorber" or, more simply, "absorber" are, except for a special case, adopted in this book. The EM-wave absorber refers to structures that can absorb an incident EM wave based on the principles of transforming the incident EM-wave energy into Joule heat or canceling mutually the phases between the incident EM wave and the reflected wave.

An object that completely absorbs all light wavelengths is known as a black body, and carbon is considered as nearly a black body. As for sound wave environments, sound-absorbing materials have been often utilized, and glass fibers, rock wools, etc. have been used as materials that absorb sound waves well. Thus, even before the EM-wave absorber was developed, objects that can be referred to as "absorbers" have been used in various scenarios in our daily lives.

The study of wave absorbers is said to date back to the study of EM-wave absorbers for the 2-GHz band carried out in the mid-1930s at the Naamlooze Vennootschap Machinerieen in the Netherlands [2].

Ever since the various types of EM-wave absorbers were developed, mostly for anechoic chamber applications, they basically have been composed of carbon-based materials.

During World War II, research began to be carried out, associated with the deep interest in wave absorbers for military use. For example, in the German Schornsteinfeger Project, two types of wave absorbers used for radar camouflage by mounting them on the periscope and snorkel of a submarine were developed [3]. One of the wave absorbers was made of a material called "Wesch," in which a carbonyl iron material is dispersed in a rubber sheet. The other, namely, the Jaumann absorber [3], was one in which a resistance sheet and a dielectric (plastic plate) were alternately superimposed, as shown in Figure 1.1d in Section 1.3. In addition, in the United States, in a project organized by O. Halpern at the MIT Radiation Laboratory, with the aim of realizing a coating-type wave absorber, the Halpern antiradar paint (HARP) was developed.

This was an EM-wave absorber using an artificial dielectric with a thickness of approximately 0.6 mm. It had a high-performance wave absorber with a resonance characteristic at the X-band. Furthermore, the "Salisbury screen absorber" was also developed at the same time in the Radiation Laboratory [4]. This was a resonant-type wave absorber, as shown in Figure 1.1b, and its structure was composed of the resistive sheet with a resistance value of 377 Ω, which was placed in a location $\lambda/4$ away from the back conductor plate.

In addition, from a practical standpoint, such as for performing measurements related to electronic devices and antenna characteristics, there is a need for an anechoic chamber. For this countermeasure, a pyramidal wave absorber capable of absorbing broadband EM waves was developed by Neher et al. in 1953. Owing to the development of this kind of a wave absorber, high accuracy has been achieved in experiments such as in the measurement of antenna radiation patterns in an anechoic chamber [5].

From a theoretical approach viewpoint, scattering waves from a planar multilayer absorber and a wedge-type absorber aimed at use for broadband wave absorbers for anechoic chambers were analyzed. This kind of analysis was conducted by G. Franceschetti and colleagues, who introduced an approximate analysis method of Riccati differential equations and optical approximation [6].

Currently, as wave absorbers based on new concepts, autonomously controllable wave absorbers [7, 8] have been promoted aggressively. In addition, wave absorbers based on the idea of a left-handed metamaterial [9] have been proposed.

In the next section, the EM-wave absorber is explained in detail from various viewpoints.

1.2 Fundamentals of Absorber Characteristics

The ideal wave absorber is able to absorb all incident EM-waves, regardless of the incident wave direction, polarization, and frequency. In other words, it is an object that does not cause any reflection waves. In practice, however, an ideal EM-wave absorber does not exist. Therefore, the performance of EM-wave-absorbing characteristics has been defined by the method of providing beforehand the allowable value assigned as the reflection coefficients. Usually, the reflection coefficient is defined to be −20 dB or less; when high performance is required, it is assumed to be −30 dB or less, as shown in Table 1.1 [10].

As the quantitative value that indicates the EM absorption performance, the reflection coefficient is represented generally in decibels. This reflection coefficient can be also regarded as return loss. A value of −20 dB corresponds

Table 1.1 Representations of reflection coefficient in wave absorbers.

Reflection coefficient (dB)	Electric-field reflection coefficient (S)	Electric-field standing-wave ratio (VSWR)	Power reflection coefficient[a),b)]
−20	0.1	1.2	0.01
−30	0.03	1.06	0.001

a) −20 dB means that 99% of the EM-wave energy incident on the absorber is absorbed if converting to energy.
b) At −30 dB, 99.9% of energy is absorbed.

to an electric-field reflection coefficient of 0.1 or a power reflection coefficient of 0.01.

From an energy viewpoint, a value of −20 dB means also that 99% of the EM-wave energy that is incident on the wave absorber is absorbed. Also, for a reflection coefficient of −30 dB, 99.9% of the incident EM-wave energy to the EM-wave absorber is absorbed. Conventionally, the absorption amount of the EM-wave absorber was evaluated using a voltage standing-wave ratio (VSWR). Recently, the reflection coefficient or return loss mentioned earlier has been used. For the VSWR value, −20 dB is equivalent to 1.2. As a special case, in the wireless local area network (LAN) field, which has led to increased demand for wave absorbers, the acceptable reflection coefficient is regarded as −6 dB or less.

1.3 Classifications of Absorbers

As is well known, EM-wave absorbers are classified according to various factors such as the structure, the material to be used, and the frequency band to be applied, as listed in Table 1.2. In this section, the wave absorbers related to an incident wave radiated from a far oscillator are explained – that is, the absorber against a plane wave case.

1.3.1 Classifications by Appearance

First, let us classify the EM-wave absorbers from their appearance [10]. There are various types of EM-wave absorbers, such as those shown in Figure 1.1.

1.3.1.1 Single-layer-type Absorber

As shown in Figure 1.1a, a flat-plate-type wave absorber is composed of a structure in which the wave absorber surface against the normal incident EM-wave direction is flat. A typical example of this type of wave absorber is a ferrite wave absorber.

Table 1.2 Classifications of wave absorbers.

Classification	Item	Remarks
1. Material classification		
	1) Conductive material 2) Dielectric material 3) Magnetic material 4) Metamaterial 5) Special material	1) Carbon materials such as carbon black or graphite have become a major material. Also, metal-based material, or the like, having a resistance is used 2) Carbon rubber, carbon-containing foamed urethane, and carbon-containing foamed polystyrene, which are made by mixing carbon into rubber or urethane 3) Ferrite or carbonyl iron material is used mainly 4) Metamaterials called as left-handed are used 5) Special material
	a) Materials based on equivalent transformation method of material constant b) Substrate-type material mounting an integrated circuit c) Substrate-type material equipped with autonomous-control-type circuit	a) By means of combinations or modifications of existing materials, the wave absorber materials create new characteristics b) Wave-absorbing material consisting of a microwave integrated circuit substrate c) Material composed of active elements, sensors, and a microchip computer, on the same substrate
2. Classification by configuration form		
(I) Classification from the number of layers	1) Single-layer-type wave absorber 2) Two-layer-type wave absorber 3) Multilayered wave absorber	1) EM-wave absorber which is made from a single layer 2) EM-wave absorber which consists of two layers having different material constants 3) Wave absorber which is constituted of three or more layers

(II) Classification by shape	1) Flat plate-shaped wave	1) Flat configuration of radio wave incident surface
	2) Quarter-wavelength wave absorber	2) Wave absorber having the configuration where the film-shaped resistor is placed in quarter wavelength apart from a conductive plate
	3) Multilayered wave absorber	3) Wave absorber having configurations of different layered material constants
	4) Jaumann absorber	4) Wave absorber superimposing alternating resistive sheet and the dielectric plate
	5) Chevron-shaped wave	5) Wave absorber composed of chevron shape at radio wave incident surface
	6) Pyramidal wave absorber	6) Wave absorber composed of tapered pyramidal structure at incident side
(III) Classification by frequency characteristics	1) Narrowband-shaped wave absorber	1) Wave absorber having the fractional bandwidth $f/f_0 = 10$–20% approximately.
	2) Broadband-type wave absorber	2) P value is more than 20%, and a wave absorber having peak or twin peaks.
	3) Ultra-wideband-shaped wave absorber	3) In more than a certain lower limit frequency, the wave absorber shows an allowable reflection attenuation characteristic or less

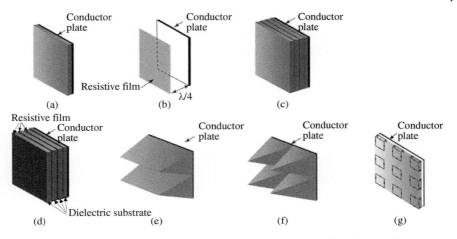

Figure 1.1 Main classifications of the wave absorber. (a) Plane type, (b) $\lambda/4$ type, (c) multilayer type, (d) Jaumann absorber, (e) sawtooth type, (f) pyramidal type, and (g) metamaterial type.

1.3.1.2 Quarter-wavelength-type Absorber

A quarter-wave-type wave absorber is constructed by placing a conductor plate in a position a quarter wavelength away from the film-shaped resistor, as shown in Figure 1.1b.

1.3.1.3 Multilayered Absorber

As illustrated in Figure 1.1c, multilayered wave absorbers are constructed by layering the absorbing materials to obtain the matching characteristic by adjusting each input impedance for each material stepwise.

1.3.1.4 Jaumann Absorber

As shown in Figure 1.1d, a Jaumann absorber consists of a configuration in which alternating resistive sheets and dielectric plates are superimposed.

1.3.1.5 Sawtooth-shape Absorber

This EM-wave absorber surface is a sawtooth type and has a kind of tapered shape, as shown in Figure 1.1e. Because of this configuration, this absorber shape has been called a chevron-shaped absorber. Although this is a single-polarization-type EM-wave absorber, it becomes possible to absorb EM waves over a wide frequency band efficiently, as with the pyramid type, which is treated next.

1.3.1.6 Pyramidal Wave Absorber

As shown in Figure 1.1d, because the pyramidal wave absorber adopts a pyramidal shape from the EM-wave incident side, this absorber exhibits EM-wave absorption characteristics over a wide frequency band to both polarized EM waves. This wave absorber is made by impregnating urethane foam, Styrofoam, or the like with a carbon material. This absorber has been widely put into practical use.

1.3.1.7 Absorbers by Artificial Materials and Special Materials

Recently, as described later, wave absorbers related to left-handed metamaterials have been proposed.

1.3.2 Classifications of Material

1.3.2.1 Conductive Absorber Material

Wave-absorbing materials that have been used since the wave absorber was invented include lossy conductive metal materials, resistive powders, and resistive films. These can be said to be typical EM-wave absorbers. This is because they are based on the principle of changing the currents generated in the absorber by the incident wave into Joule heat. As conductive wave-absorbing materials, there exist materials having predetermined resistance values. These are composed mainly of carbon-based materials such as carbon black or graphite. They are widely used in the form of platelike or filmlike materials for the conductive type of wave absorber material. Furthermore, excellent EM-wave-absorbing characteristics are realized if using a specific conductive fabric. A typical example of an EM-wave absorber using a resistive conductive material is a $\lambda/4$-type wave absorber, which is a basic EM-wave absorber configuration.

1.3.2.2 Dielectric Absorber Material

Examples of dielectric wave-absorbing materials are carbon rubber, carbon-containing urethane foam, and carbon-containing expanded polystyrene. These materials are made by mixing carbon material with rubber, urethane, etc. This kind of material is used to realize broadband absorption characteristics and is applied to multilayer-structure, wedge, or pyramid types of wave absorbers, as described earlier.

1.3.2.3 Magnetic Absorber Material

A thin wave absorber configuration can be realized using ferrite, carbonyl iron, and the like, which are magnetic loss materials available at frequencies higher than the very high frequency (VHF) band. In this case, the EM-wave-absorbing characteristic is strongly governed by the frequency dispersion characteristic of magnetic material and, thus, by permeability value.

1.3.2.4 Metamaterial

Recently, wave absorbers have been proposed as one of the applications of the metamaterial that is called "left-handed." Exploiting the idea of the left-handed metamaterial has made possible new types of absorbers that do not require a back conductor plate [9], and terahertz band absorbers have been suggested. Further, EM-wave absorbers based on a novel configuration concept have been proposed, and they are summarized as follows.

(a) To realize new EM-wave-absorption characteristics, an absorber based on the idea of equivalently converting material constants by means of loading some kind of metal pattern on an existing material surface or making small holes, and the like has been proposed (see Chapter 9). These methods are unified as the "equivalent transformation method of material constants." By introducing this concept, wave absorbers much thinner than the conventional ones can be realized [11, 12].

(b) A new wave absorber is a type composed of a microwave integrated circuit. This wave absorber has a simple, yet lightweight, structure, and the broadband-absorbing characteristics can be realized effectively, even beyond the microwave frequencies [13, 14].

(c) An autonomously controllable metamaterial-type wave absorber is a wave absorber based on a completely new material concept; thus, its structure is composed of a type of artificial material that is equipped with the active element circuit, sensors, and microchip computer on the same substrate [7, 8, 15–17].

The concept of this material configuration is based on the autonomy of living tissue, as is described in Chapter 10.

1.3.3 Classifications by Configuration Forms

Furthermore, the wave absorber is categorized from the viewpoint of the "number of layers" constituting each absorber layer and the "shape of appearance" in the absorber structure.

1.3.3.1 Classification from Layered Numbers

(a) Single-layer-type absorber
An absorber made from a single-layer material is called a "single-layer-type wave absorber." Normally, a metal plate made of aluminum, copper, or the like is attached to the back of an absorber. This type of EM-wave absorber can be seen in those using ferrite, carbonyl iron material, and other such materials.

(b) Two-layer-type absorber
This is a wave absorber that has two layers composed of different material constants. This configuration is often introduced when aiming at improving a single-layered absorber's characteristic to realize a more broadband absorber characteristic.

(c) Multilayered absorber

The multilayered wave absorber is usually a wave absorber consisting of three or more layers. In the multilayered wave absorber, the wideband characteristics are obtained by increasing the number of layers, and this kind of absorber can be used, for example, for an anechoic chamber.

1.3.4 Classifications by Frequency Characteristics

Regarding the quality of the EM-wave absorber characteristic, the "goodness of absorption characteristic" is defined by introducing the idea of a "figure of merit [10]."

For example, when evaluating a reflection coefficient below $-20\,\text{dB}$ as a good EM-wave absorber, if the bandwidth cut by the level of $-20\,\text{dB}$ values is assumed to be Δf, by dividing the bandwidth values Δf with the center frequency f_0, the figure of merit can be defined as $\Delta f/f_0$. This characteristic is mainly classified into the three types shown in Figure 1.2.

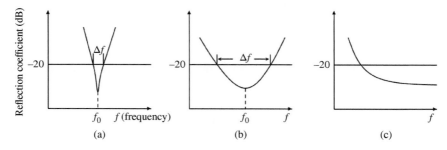

Figure 1.2 Classification by frequency band. (a) Narrowband type, $\Delta f/f_0 \times 100 = 10-20\%$; (b) broadband type $\Delta f/f_0 \times 100 = 20-30\%$; and (c) ultra-wideband type $\Delta f/f_0 \times 100 = 30\%$ and above [10].

1.3.4.1 Narrowband-type Absorber

This is usually associated with the case of the characteristic that can be found in a single-layer wave absorber, or the like. In this case, if the figure of merit is expressed as a percentage, it is approximately 10–20%, as illustrated in Figure 1.2a. When a narrow frequency band is needed, as in the case of a radar application, this type of absorber is used.

1.3.4.2 Broadband-type Absorber

Notice that the distinction between the case of the wideband and the narrowband types is not clearly defined. In the case where the percentage of $\Delta f/f_0$ is not less than 20%, the EM-wave absorbers often show a peak or twin-peak characteristic, as shown in Figure 1.2b. These cases are generally referred to as "broadband wave absorbers."

1.3.4.3 Ultra-wideband-type Absorber

This type of absorber possesses a wideband-absorbing characteristic in the above-assigned absorbing frequency, which is set beforehand. This results in an absorption characteristic of a type in which the lower limit frequency of the allowable reflection coefficient to the EM-wave absorber meets the specified frequency.

In other words, it is an EM-wave absorber with broadband characteristics that can absorb EM waves above a frequency determined beforehand, as depicted in Figure 1.2c. Of course, because $\Delta f / f_0$ becomes infinite, the definition of the figure of merit cannot be used in this case. In general, the multilayer absorber, wave absorber of the saw-tooth shape, and pyramidal shape exhibit this kind of property.

1.4 Application Examples of Wave Absorbers

The examples of the main application of the wave absorber and the related materials used therein are given in Table 1.3. As shown in Table 1.3, the EM-wave absorber application fields are expanding along with the development of communication technologies.

As shown in Table 1.3, with respect to the wave absorbers of the anechoic chamber described previously, many studies have been conducted, and much research that incorporates the latest analysis has been published. For example, with respect to the pyramid-type or wedge absorbers, research based on theory and experiments [18], the moment method [19, 20], and the frequency-domain finite-difference method [21] have been reported. Further, careful experimental studies have been conducted to study the anechoic chamber [22–24].

In addition, as one of the main application fields of the EM-wave absorber, the topic of radar technology improvement has been examined. Particularly, applications of the EM-wave absorber associated with radar problems have been studied from the early stages of wave absorber development [25–30]. Countermeasures to the problem of false images that occur on a radar screen, caused by radar waves reflected from a large bridge over the strait, have been in demand. For high-rise buildings, wave absorber walls were developed in Japan, because the reflection of the TV EM wave from a building wall surface causes a TV ghost (see Chapter 8). Furthermore, from the viewpoint of EM environment conservation in various consumer electronics products and wireless LAN environments, EM-wave leakage measures have been considered, leading to the development of various types of wave absorbers. In addition, measures to reduce noises generated from printed circuit boards are needed from the standpoint of absorber material and are an issue in recent EM compatibility research.

As for these countermeasures, a thin plate-shaped sheet made of a magnetic material called a noise-suppressing element, fine ferrite beads, etc. has been

Table 1.3 Examples of main wave absorber use.

Application examples	EM-wave absorber, and material used
For anechoic chamber (more than 30 MHz)	• Wave absorber of combination multilayer structure with carbon-based material and ferrite • Pyramid-type wave absorber material being produced by mixing carbon in urethane foam material, or wave absorber in which sawtooth type unit absorbers made of a carbon material are arranged alternately in vertical and horizontal directions. • Ferrite single-layer wave absorber (simplified type) • Electromagnetic-wave absorber material that is composed of a combination of a dielectric comprising metal fiber material and a ferrite
Improvement of radar characteristics	• Absorber material using sintered ferrite • Absorber material of rubber ferrite • Absorber material composed of a nonwoven fabric and metal fibers
For high-rise building wall (TV ghost prevention measures, 100 MHz – an example of the old analog broadcasting)	• Absorber material of ferrite tile • Absorber material using ferrite and dielectric combination • Absorber material mixing ferrite grains into concrete
For electromagnetic interference prevention (for prevention of leakage wave of a microwave oven, wireless LAN measures) (2.45 and 5.2 GHz)	• Wave absorber using rubber ferrite • Wave absorber using resin ferrite • Wave absorber composed of carbon-based dielectric material and building materials • Wave absorber composed of resistance film-based materials and building materials • Wave absorber composed of ferrite and building materials
Countermeasure for electronic circuit noise (10 MHz to 5 GHz)	• Sheets composed of special magnetic materials and electrically conductive material • Insulation sheet which has ferrite powder mixed with polymer • Composite material made from metallic flat powder • Small cylindrical-shaped ferrite
For mobile communication (malfunction prevention measures of electronic automatic billing system, 5.8 GHz)	• Wave absorber material consisting mainly of ferrite • Foam material containing a conductive material • Wave absorber material containing metal fiber in nonwoven fabric • Wave absorber material coated with a conductive paint on synthetic fibers • Paved road wave absorption material consisting of carbon fiber and asphalt material

developed. Furthermore, malfunction prevention measures based on EM-wave scattering in the site of an electronic automatic billing system are an example of the application of EM-wave absorbers for mobile communications.

Moreover, EM-wave absorbers are expected to play an important role also from the viewpoint of communication control in the automated driving vehicle technology, which has been rapidly developing recently, and in infrastructure development related to this area of research. Thus, the application fields of EM-wave absorbers are expanding in response to the recent development of communication technology, as shown in Table 1.3.

References

1 Vandrer Vorst, A., Rosen, A., and Kotsuka, Y. (2006). *RF/Microwave Interaction with Biological Tissue*. Wiley Interscience.

2 Naamlooze Vennootschap Machinerieen, French Patent 802 728, Feb. 19, 1936.

3 Schade, H.A. (1945). Schornsteinfeger. U.S. Tech. Mission to Europe, *Tech. Rep. 90-45 AD-47746*.

4 W. W. Salisbury, "Absorbent body for electromagnetic waves," US patent 2599944, filed May 11, 1943, granted Jun. 10, 1952.

5 Emerson, W. (1973). Electromagnetic wave absorbers and anechoic chambers through the years. *IEEE Trans. Antennas Propag.* 21 (4): 484–489.

6 Franceschetti, G. (1964). Scattering from plane layered media. *IEEE Trans. Antennas Propag.* 12: 754–763.

7 Kotsuka, Y. and Amano, M. (2004). New concept for functional electromagnetic cell material for microwave and millimeter use. *IEEE MTT-S International Microwave Symposium Digest*.

8 Kotsuka, Y., Murano, K., Amano, M., and Sugiyama, S. (2010). Novel right-handed metamaterial based on the concept of autonomous control system of living cells and its absorber applications. *IEEE EMC Trans.* 52 (3): 556–565.

9 Landy, N.I., Sajuyigbe, S., Mock, J.J. et al. (2008). Perfect metamaterial absorber. *Phys. Rev. Lett.* 100: 207402.

10 Shimizu, Y. (Editorial Committee Chairman) (1999). *Electromagnetic Waves Absorption and Shielding*, 123. Tokyo: Nikkei Gijyutu Tosho.

11 Amano, M. and Kotsuka, Y. (2003). A method of effective use of ferrite for microwave absorber. *IEEE Trans. Microwave Theory Tech.* 51 (1): 238–245.

12 Amano, M. and Kotsuka, Y. (2015). Detailed investigations on flat single layer selective magnetic absorber based on the equivalent transformation method of material constants. *IEEE Trans. Electromagn. Compat.* 57 (6): 1398–1407.

13 Kotsuka, Y. and Amano, M. (2003). Broadband EM-wave absorber based on integrated circuit concept. In: *IEEE MTT-S International Microwave Symposium Digest*, vol. 2, 1263–1266.

14 Kotsuka, Y. and Kawamura, C. (2005). Proposal of a new EM-wave absorber based on integrated circuit concept. *IEICE Trans.* J88-C (12): 1142–1148.

15 Kotsuka, Y. and Amano, M. (2003). Microwave Functional Material for EMC. *Tech. Rep. of IEICE, EMCJ 3003-40*, pp. 13–18.

16 Kotsuka, Y. and Amano, M. (2004). A new EM-wave absorber using functional electromagnetic cell material. In: *Proceedings of the EMC'04 Sendai*, vol. 1, 301–304.

17 Kotsuka, Y., Sugiyama, S., and Kawamura, C. (2007). Novel computer control metamaterial beyond conventional configuration and its microwave absorber application. In: *IEEE MTT-S International Microwave Symposium Digest*, 1627–1630.

18 Dewitt, B.T. and Burnsid, W.D. (1988). Electromagnetic scattering by pyramidal and wedge absorber. *IEEE Trans. Antennas Propag.* 36 (7): 971–984.

19 Yang, C.F., Burnside, W.D., and Rudduck, R.C. (1992). A periodic moment method solution for TM scattering from lossy dielectric bodies with application to wedge absorber. *IEEE Trans. Antennas Propag.* 40 (9): 652–660.

20 Yang, C.F., Burnside, W., and Rudduck, R.C. (1993). A doubly periodic moment method solution for the analysis and design of an absorber covered wall. *IEEE Trans. Antennas Propag.* 41 (5): 600–609.

21 Sun, W., Liu, K., and Balanis, C.A. (1996). Analysis of singly and doubly periodic absorbers by frequency-domain finite-difference method. *IEEE Trans. Antennas Propag.* 44 (6): 798–805.

22 Tofani, S., Ondrejka, A.R., and Hill, D.A. (1991). Bistatic scattering of absorbing materials from 30 to 1000 MHz. *IEEE Trans. Electromagn. Compat.* 34 (3): 304–307.

23 Tofani, S., Ondrejka, A., and Kanda, M. (1991). A time-domain method for characterizing the reflection coefficient of absorbing materials from 30 to 1000 MHz. *IEEE Trans. Electromagn. Compat.* 33 (3): 234–240.

24 Johnk, R.T., Ondrejka, A., Tofani, S., and Knada, M. (1993). Time-domain measurement of the electromagnetic backscatter of pyramidal absorber and metallic plates. *IEEE Trans. Electromagn. Compat.* 35 (4): 429–433.

25 Bhattacharyya, A.K. (1989). Electromagnetic scattering from a flat plate with rim and RAM saving. *IEEE Trans. Antennas Propag.* 37 (5): 659–663.

26 Wong, P.T.C., Chambers, B., Anderson, A.P., and Wright, P.V. (1992). Large area conducting polymer composites and use in microwave absorbing material. *Electron. Lett.* 28 (17): 289–290.

27 Chambers, B. (1995). Symmetrical radar absorbing structures. *Electron. Lett.* 31 (5): 404–405.

28 Knott, E.F. (1997). Suppression of edge scattering with impedance strings. *IEEE Trans. Antennas Propag.* 45 (12): 1768–1773.

29 Williams, T.C., Stuchly, M.A., and Saville, P. (2001). Modified
transmission-reflection method for measuring constitutive parameters of
thin flexible high-loss materials. *IEEE Trans. Microwave Theory Tech.* 51 (5):
1560–1566.

30 Matous, K. and Dvorak, G.J. (2003). Optimization of electromagnetic
absorption in laminated composite plates. *IEEE Trans. Magn.* 39 (3):
1827–1835.

2

Fundamental Theory of EM-Wave Absorbers

In this chapter, in order to understand electromagnetic (EM)-wave absorbers, fundamental theories are described in detail from the viewpoints of both transmission line (TL) theory and Maxwell's equations. A plane EM-wave problem is often treated by replacing it with a TL problem that is equivalent in its plane-wave behavior. With this approach, it becomes easy to understand the performance of some EM-wave devices, the physical meaning, and other aspects without solving Maxwell's equations as to transverse electromagnetic (TEM)-wave problems. Even in the case of EM-wave absorber problems, they often treat the TL problems in order to understand the operating principles and absorption characteristics.

Section 2.1 describes basic concepts and theories, such as the reflection coefficient, impedance related to TL-line, which are necessary for understanding wave absorber characteristics from the viewpoint of TL theory.

Section 2.2 clarifies in detail the fundamental constitutive theory of the Smith chart and the procedures for applying it.

In Section 2.3, Maxwell's equations are presented, including their derivation methods.

Next, reflection from a perfect conductor surface and its standing-wave behavior are explained from an EM-wave theory viewpoint.

Further in Section 2.3, the behaviors of normal and oblique incident waves are described at the interface between the two media expanding to infinity.

Finally, as the basic knowledge necessary for designing EM-wave absorbers, the fundamental theory of multiple reflections is also taken up.

2.1 Transmission Line Theory

From the example of a parallel two-wire feeder used for the early TV receiver antenna, one can easily imagine that two parallel conducting wires can transmit EM waves. Microwave waveguides, coaxial waveguides, many types of microwave devices, etc., can often be treated equivalently on the basis

Electromagnetic Wave Absorbers: Detailed Theories and Applications, First Edition. Youji Kotsuka.
© 2019 John Wiley & Sons, Inc. Published 2019 by John Wiley & Sons, Inc.

of the TL theory. Even the EM-wave absorber problems are so often easily understood by an equivalent replacement by the TL theory [1–3]. For these reasons, let us first describe the microwave TL theory in detail.

When the EM-wave power is transmitted along a TL, such as two parallel wires, a lossless TL composed of inductance and capacitance based on both the magnetic lines of force and the electric lines of force generated around them must be constructed. But note that these inductance and capacitance cannot be considered as just one concentrated circuit element in the same way as for low-frequency circuits.

An electric circuit in which each element is distributed in this manner throughout the TL is called a "distributed constant circuit."

In the following, the telegraph equations are first derived, and then the definition of impedance and the matching condition for TL based on this are described.

2.1.1 Transmission Line Equation

Figure 2.1a shows a high-frequency TL. Figure 2.1b illustrates an enlarged microsection between P and Q on its TL in Figure 2.1a.

In this TL, let us denote the inductance, capacitance, resistance, and conductance per unit length using L [H/m], C [F/m], R [1/m], and G [S/m], respectively.

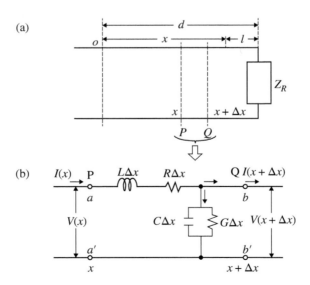

Figure 2.1 Equivalent circuit representation of the electromagnetic-wave transmission line. (a) Depicts the terminal part, including the load on the transmission line. (b) General circuit representation of the transmission line, including loss.

Therefore, when denoting these electrical constant values on the TL, Δx has to be multiplied; and they can be expressed as shown in Figure 2.1b.

When considering the current in section Δx in Figure 2.1b, that is, $aa' - bb'$, one can find that the following relationship holds:

(Current of the input terminal aa')

= (Current flowing to the output terminal bb')

+ (Branch current flows to $C\Delta x$ and $G\Delta x$).

These relationships can be expressed as

$$I(x) = I(x + \Delta x) + (j\omega C\Delta x + G\Delta x)V(x + \Delta x).$$

Dividing both sides in the given expression by Δx,

$$\frac{I(x + \Delta x) - I(x)}{\Delta x} = -(j\omega C + G)V(x + \Delta x).$$

Taking the limit of $\Delta x \to 0$, the following equation for the current can be obtained:

$$\frac{dI(x)}{dx} = -(G + j\omega C)V(x) = -Y(x)V(x), \tag{2.1}$$

where $Y = G + jC$.

Similarly, paying attention to the voltage between $aa' - bb'$ in Figure 2.1b, the following relation holds for it:

(Voltage at the input terminal aa') = (Voltage at the output terminal bb')

+ (Voltage drop due to $R\Delta x$ and $L\Delta x$)

$$V(x) = V(x + \Delta x) + (j\omega L\Delta x + R\Delta x)I(x).$$

From this relation,

$$\frac{V(x + \Delta x) - V(x)}{\Delta x} = -(j\omega L + R)I(x).$$

Considering the limit of $\Delta x \to 0$, the following voltage equation is obtained:

$$\frac{dV(x)}{dx} = -(R + j\omega L)I(x) = -ZI(x), \tag{2.2}$$

where $Z = R + j\omega L$.

Equations (2.1) and (2.2) are called the telegraphist's equations. Then, differentiating Eq. (2.2) with respect to x, and substituting Eq. (2.1),

$$\frac{dV^2(x)}{dx^2} = -Z\frac{dI}{dx} = ZYV(x) = \gamma^2 V(x). \tag{2.3}$$

Similarly, by differentiating Eq. (2.1) with respect to x, and substituting Eq. (2.2),

$$\frac{dI^2(x)}{dx^2} = YZI(x) = \gamma^2 I(x). \tag{2.4}$$

Here,

$$\gamma^2 = ZY = (R + j\omega L)(G + j\omega C). \tag{2.5}$$

Hence,

$$\gamma = \sqrt{(R + j\omega L)(G + j\omega C)} = \alpha + j\beta. \tag{2.6}$$

Here, γ is a propagation constant, ω is an angular frequency, the real part α represents the attenuation constant, and the imaginary part β denotes the phase constant. These differential Eqs. (2.3) and (2.4) formally have the same form. Accordingly, their solutions are generally given by a trigonometric, exponential, or hyperbolic function. In this case, the solution that should be selected must be determined taking into account the physical phenomena under consideration. Since we first consider the lossless TL, we take both exponential functions, $e^{-\gamma x}$ and $e^{\gamma x}$ as the solutions of both Eqs. (2.3) and (2.4).

Also, since $R = 0$ and $G = 0$, the propagation constant, impedance, and admittance are expressed by the following equation using the angular frequency ω, respectively.

$$\gamma = j\omega\sqrt{LC} = j\beta, \quad Z = j\omega L, \quad Y = j\omega C.$$

Accordingly, the solution to the voltage equation of Eq. (2.3) can be chosen in the following form using arbitrary constant values V_1 and V_2:

$$V(x) = V_1 e^{-\gamma x} + V_2 e^{\gamma x}$$

On the other hand, from the relation of Eq. (2.2), the solution for the electric current $I(x)$ is given by the following equation for the lossless TL:

$$\frac{dV(x)}{dx} = -\gamma V_1 e^{-\gamma x} + \gamma V_2 e^{\gamma x} = -ZI(x).$$

Hence,

$$I(x) = \frac{\gamma}{Z}(V_1 e^{-\gamma x} - V_2 e^{\gamma x})$$

$$= \frac{1}{Z_c}(V_1 e^{-\gamma x} - V_2 e^{\gamma x}),$$

where Z_c is called the characteristic impedance in the case of a lossless line, and is expressed by the following equation:

$$Z_c = \sqrt{Z/Y} = \sqrt{L/C} = R_c.$$

That is,

$$V(x) = V_1 e^{-\gamma x} + V_2 e^{\gamma x} \tag{2.7a}$$

$$I(x) = \frac{1}{Z_c}(V_1 e^{-\gamma x} - V_2 e^{\gamma x}). \tag{2.7b}$$

The impedance Z_c is equal to the "characteristic resistance R_c" of the considered TL. Using these relationships, the equation for the current $I(x)$, which needs to be solved, takes the following form:

$$I(x) = \frac{1}{\sqrt{L/C}}(V_1 e^{-\gamma x} - V_2 e^{\gamma x}) = \frac{1}{R_c}(V_1 e^{-\gamma x} - V_2 e^{\gamma x}). \tag{2.8}$$

It should be noted here that the expressions derived so far have been expressed at an arbitrary point x on the TL, starting from the transmitter side and ending at the load terminal side.

Now, let us focus on the load end to determine the values of V_1 and V_2.

Assuming that the load impedance at the load end, shown in Figure 2.1, is Z_R, and the voltage and current are expressed by $V(d) = V_R$ and $I(d) = I_R$, respectively, one has

$$V_R = Z_R I_R.$$

Therefore, considering the load end $x = d$ in Eqs. (2.7a) and (2.7b) and (2.8),

$$V_R = Z_R I_R = V_1 e^{-\gamma d} + V_2 e^{\gamma d}$$
$$R_c I_R = V_1 e^{-\gamma d} - V_2 e^{\gamma d}.$$

By taking the sum and difference in both of these equations, the values of V_1 and V_2 at the load end of the terminal can be determined:

$$\left. \begin{array}{l} V_1 = (Z_R + R_c)\frac{I_R}{2}e^{\gamma d} \\ V_2 = (Z_R - R_c)\frac{I_R}{2}e^{-\gamma d} \end{array} \right\}. \tag{2.9}$$

Furthermore, a relationship $d - x = l$ is introduced to change the reference point of the load terminal end into the any point on TL. By doing so, the variable x, expressed by two exponential functions in Eqs. (2.7a) and (2.7b), can be changed to the variable l. Substituting the relation $d - x = l$ into Eqs. (2.7a) and (2.7b), the following equation can be derived:

$$V_l = (Z_R + R_c)\frac{I_R}{2}e^{\gamma d}e^{-\gamma(d-l)} + (Z_R - R_c)\frac{I_R}{2}e^{-\gamma d}e^{\gamma(d-l)} = V_i e^{\gamma l} + V_r e^{-\gamma l}$$

$$I_l R_c = (Z_R + R_c)\frac{I_R}{2}e^{\gamma d}e^{-\gamma(d-l)} - (Z_R - R_c)\frac{I_R}{2}e^{-\gamma d}e^{\gamma(d-l)}$$

$$= V_i e^{\gamma l} - V_r e^{-\gamma l}.$$

Since the TL is now lossless, replacing the propagation constant by

$$\gamma = j\omega\sqrt{LC} = j\beta, \quad \beta = 2\pi/\lambda : \text{wavelength of EM-wave.}$$

From these expressions,

$$\left.\begin{array}{l} V_l = V_i e^{j\beta l} + V_r e^{-j\beta l} \\ I_l R_c = V_i e^{j\beta l} - V_r e^{-j\beta l} \end{array}\right\}, \tag{2.10}$$

where

$$V_i = (Z_R + R_c)\frac{I_R}{2}, \quad V_R = (Z_R - R_c)\frac{I_R}{2}. \tag{2.11}$$

Further, when rewriting Eq. (2.10), using the Euler's formula $e^{\pm j\beta l} = \cos\beta l \pm j\sin\beta l$,

$$V_l = V_R \cos\beta l + jR_c I_R \sin\beta l$$
$$R_c I_l = R_c I_R \cos\beta l + jV_R \sin\beta l. \tag{2.12}$$

In this way, the voltage V_l and current I_l at the distance l from the load terminal end in the direction of transmitter are derived. According to this equation, the impedance Z_l at the distance l from the load terminal end along the TL can be presented by the following equation:

$$Z_l = \frac{V_l}{I_l} = R_c\frac{V_R \cos\beta l + jR_c I_R \sin\beta l}{R_c I_R \cos\beta l + jV_R \sin\beta l}. \tag{2.13}$$

After rearrangement of Eq. (2.13), focusing on $I_R \cos\beta l$, it reduces to

$$Z_l = R_c\frac{Z_R + jR_c \tan\beta l}{R_c + jZ_R \tan\beta l}. \tag{2.14}$$

Further, by normalizing Eq. (2.14) in terms of R_c, the following equation for the normalized impedance z_l is obtained:

$$z_l = \frac{Z_l}{R_c} = \frac{z_R + j\tan\beta l}{1 + jz_R \tan\beta l}, \tag{2.15}$$

where $z_R = \dfrac{Z_R}{R_c}$.

Now, R_c stands for the characteristic resistance of the TL. Summarized here are cases where the load impedance takes special values in Eq. (2.15).

(a) The case where the TL end is shorted ($Z_R = 0$),

$$z_l = \frac{Z_l}{R_c} = j\tan\beta l. \tag{2.16}$$

(b) The case where the TL end is opened ($Z_R = \infty$),

$$z_l = \frac{Z_l}{R_c} = j\cot\beta l. \tag{2.17}$$

(c) The case of $Z_R = R_c$,

$$z_l = \frac{Z_l}{R_c} = 1 \quad \therefore Z_l = R_c. \tag{2.18}$$

As follows from the expressions described in cases (a) and (b), when the TL end is short-circuited or open-circuited, varying the length from the TL end within a half-wave length, it becomes possible to implement arbitrary reactance values. Next, let us think about the TL of infinite length associated with the case of (c).

In an infinitely long TL, since it extends literally to infinity, there is no reflected wave. Accordingly, when taking into consideration Eq. (2.10), one obtains $V_l = V_i \, e^{j\beta l}$ and $I_l = V_i \, e^{j\beta l}/R_c$, since V_l and I_l can be represented using only the traveling waves. As shown in Figure 2.2.

Therefore,

$$Z_l = \frac{V_l}{I_l} = R_c.$$

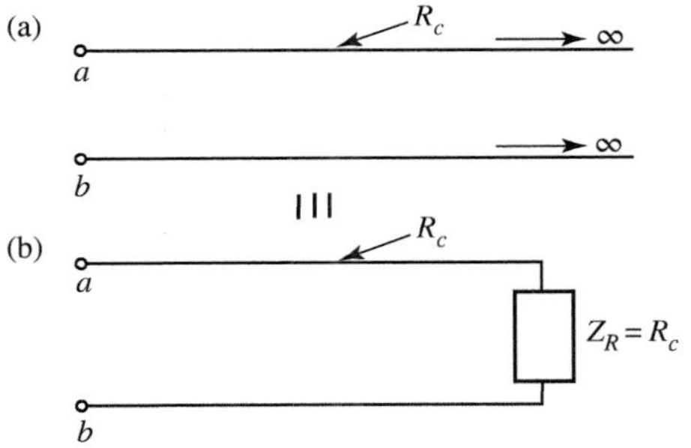

Figure 2.2 Equivalent relationship of transmission lines. (a) Infinite length transmission line. (b) A transmission line equivalent to an infinite length.

From the relationship based on the results in (c), it is found that an infinitely long line is equivalent to the terminated TL with the characteristic resistance R_c, as shown in Figure 2.2. That is, the absence of reflected waves in the TL means that the value of the characteristic impedance R_c as the load impedance Z_R must be selected. Therefore, this is the condition for taking the matching state of a TL.

2.1.2 Reflection Coefficient

2.1.2.1 Reflection Coefficient at Load Terminal End
As mentioned in the previous section, it turned out that the current I_l, voltage V_l, and impedance Z_l of the TL vary depending on the nature of the load end. Namely, the reflected wave voltage V_r relative to the incident wave voltage V_i is

determined by the nature of the load impedance. Therefore, the ratio of V_r to V_i is taken to define the reflection coefficient S_R at the load end of the TL, and the fluctuation state of the TL can be evaluated.

From Eq. (2.11),

$$S_R = \frac{V_r}{V_i} = \frac{Z_R - R_c}{Z_R + R_c} = \frac{z_R - 1}{z_R + 1} = |S_R| e^{j\varphi_R}, \tag{2.19}$$

where $z_R = Z_R/R_c$ represents the normalized impedance of the load.

Expression (2.19) defines the relationship between the reflection coefficient of the load end and the normalized load impedance. On the other hand, the normalized load impedance z_R can be represented by the following expression:

$$z_R = \frac{1 + S_R}{1 - S_R}. \tag{2.20}$$

2.1.2.2 Reflection Coefficient on Transmission Line

From a general point of view, the reflection coefficient can be determined everywhere on the TL, and not only at the load end. As shown in Figure 2.3, the impedance looking toward the load impedance Z_R at the point (a', b') away from the load end can be represented by Z_l.

This relation is in the equivalence relation to the TL connected to the load Z_l in the positions (a', b'), as shown in Figure 2.3b. Accordingly, the reflection coefficient at an arbitrary point of the TL can be defined by using Z_l, as if it

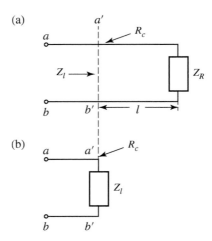

Figure 2.3 Explanation of the reflection coefficient along the transmission line.
(a) A diagram of an impedance Z_l looking into a transmission line with a load impedance Z_R at the terminal end. (b) Transmission line equivalent to the above Figure (a).

could be regarded as a load impedance. In the expression (2.19), as $Z_R \to Z_l$,

$$S_l = \frac{Z_l - R_c}{Z_l + R_c} = \frac{z_l - 1}{z_l + 1} = |S_l|e^{j\varphi_l}. \tag{2.21}$$

Then,

$$z_l = \frac{1 + S_l}{1 - S_l}. \tag{2.22}$$

Here, $z_l = Z_l/R_c$ is the normalized impedance at any point of the TL.

Consequently, according to these investigations, the following relationships between the reflection coefficient S and a normalized impedance z at the load terminal or at arbitrary points on the TL are maintained:

$$S = \frac{z - 1}{z + 1}$$

or

$$z = \frac{1 + S}{1 - S} \tag{2.23}$$

$$z = \frac{Z_R}{R_c} \text{ (load terminal end)}$$

or

$$z = \frac{Z_l}{R_c} \text{(on the transmission line).} \tag{2.24}$$

Thus, it should be noted that the reflection coefficient is closely related to the normalized impedance.

2.1.2.3 Reflection Coefficient and Standing-Wave Ratio

As is evident from Eqs. (2.7a) or (2.7b) and (2.8), if a perfect matching is not achieved, the traveling wave ($e^{-\gamma x}$) and the reflected wave ($e^{\gamma x}$) coexist at the same time on the TL, and as a result a standing wave arises. In this case, the ratio between the maximum voltage value V_{max} and the minimum voltage value V_{min}, or the ratio between the maximum current value I_{max} and the minimum current value I_{min} is defined as the "standing-wave ratio." This standing-wave ratio is used as an estimate of the reflection amount together with the reflection coefficient. When the standing-wave ratio is denoted by ρ, the reflection coefficients S and ρ are related by the following equation:

$$\rho = \frac{V_{max}}{V_{min}} = \frac{I_{max}}{I_{min}} = \frac{1 + |S|}{1 - |S|}, \tag{2.25}$$

where $|S| = |S_R| = |S_l| = \frac{\rho - 1}{\rho + 1}$.

Here, $|S_R|$ and $|S_l|$ represent the reflection coefficients at the load end and an arbitrary point on the TL, respectively. In addition, since the voltage and current are in phase with each other, when the load is viewed from the point of maximum value V_{max} or the minimum value V_{min} of the standing voltage wave, the normalized input impedance becomes a pure resistance.

In this case, the normalized resistance r that looks in the load impedance direction at the points of V_{max} and I_{min} of the standing wave is denoted by the standing-wave ratio, ρ.

On the other hand, there exists a difference that r looking in the load impedance direction at the point of I_{max} and V_{min} is given by the reciprocal of the standing-wave ratio, $1/\rho$. These relationships are shown in the following equations:

$$z_{lmax} = r_{max} = \rho \tag{2.26}$$

$$z_{lmin} = r_{min} = \frac{1}{\rho}. \tag{2.27}$$

The suffixes "*lmax*" and "*lmin*" mean the positions that look in the load impedance direction at the length l on the TL, taken in the case of V_{max} and I_{min} or I_{max} and V_{min}, respectively.

2.1.3 Transmission Line with Loss

The lossless TL which is able to transmit power without attenuation has been investigated so far, when EM-wave power is transmitted from the input terminal to the load terminal. Next, an EM-wave TL with a loss is briefly presented.

In practice, an ordinary TL has Ohmic loss as well as dielectric loss, which occurs in a dielectric material that is used as the protective material of the conductive line, although it is insignificant compared to the conductive line loss. In addition, in general, if there is no shielding in the parallel two-core wire in the TL, radiation loss occurs; and if some objects exist nearby, the loss will occur through the disturbance of the electric field. However, when considering an ideal TL based on theoretical considerations or when the line has a short length, it becomes possible to treat it as the lossless line described in the previous section.

In this section, let us briefly describe the TL with loss.

The lossy TL, as shown in Figure 2.1b, is represented by connecting a series resistor R along the TL and connecting the parallel conductance G between the pair of TLs. As for the expressions in the lossy line, the characteristic resistance R_c has to be replaced by the characteristic impedance Z_c and the propagation constant ($\gamma = j\beta$) in the lossless case has to be replaced by a propagation constant including the real part ($\gamma = \alpha + j\beta$).

That is,

$$Z_c = \sqrt{\frac{Z}{Y}} = \sqrt{\frac{R + j\omega L}{G + j\omega C}} \tag{2.28}$$

$$\gamma = \sqrt{ZY} = \sqrt{(R + j\omega L)(G + j\omega C)} = \alpha + j\beta. \tag{2.29}$$

In this way, the propagation constant for the voltage and current in the lossy TL become complex-valued numbers.

$$
\left.\begin{aligned}
V_l &= V_i e^{\gamma l} + V_r e^{-\gamma l} \\
Z_c I_l &= V_i e^{\gamma l} - V_r e^{-\gamma l} \\
V_i &= (Z_R + Z_c)(I_R/2) \\
V_r &= (Z_R - Z_c)(I_R/2)
\end{aligned}\right\}. \tag{2.30}
$$

From Eq. (2.30), the concrete equations of voltage and current in the TL, taking into account the loss, can be represented by the following equation using hyperbolic functions.

That is, applying the following relationships between hyperbolic and trigonometric functions $j \sin \beta l = \sinh j\beta l$, $j \tan \beta l = \tanh j\beta l$, and $j \tan \beta l = \tanh j\beta l$ to Eq. (2.30), the equations of voltage and current in the present case can be derived, respectively, and become, respectively,

$$
\left.\begin{aligned}
V_l &= V_R \cosh \gamma l + Z_c I_R \sinh \gamma l \\
I_l &= I_R \cosh \gamma l + (1/Z_c)V_R \sinh \gamma l
\end{aligned}\right\}. \tag{2.31}
$$

Therefore, the normalized impedance z_l looking toward the terminal load end on any point of the TL,

$$
z_l = \frac{Z_l}{Z_c} = \frac{z_R + \tanh \gamma l}{1 + z_R \tanh \gamma l}, \tag{2.32}
$$

where $z_R = Z_R/Z_c$.

2.1.4 Reflection Coefficient in Transmission Line with Loss

The reflection coefficient and the standing-wave ratio of the lossy lines are also defined in the same manner as in the case of lossless lines at the load terminal. The reflection coefficient S_R and the standing-wave ratio ρ_R at the load end are expressed by the following equation:

$$
\text{Reflection coefficient at the load point} : S_R = \frac{V_R}{V_i} = \frac{Z_R - Z_c}{Z_R + Z_c} \tag{2.33}
$$

$$
\text{Standing wave ratio near the load} : \rho_R = \frac{1 + |S_R|}{1 - |S_R|}. \tag{2.34}
$$

Unlike the lossless case, the values of the reflection coefficient and the standing-wave ratio at any point of the lossy TL take different values for each point of the TL due to line loss. At a point which moves toward the power supply side with the distance l from the load end, the reflection coefficient (S_l) and the standing-wave ratio (ρ) are represented by the following equations by replacing $j\beta$ in the lossless case with $\alpha + j\beta$:

$$S_l = S_R e^{-2\gamma l} = |S_R| e^{-2\alpha l} e^{j(\varphi_R - 2\beta l)} \tag{2.35}$$

$$\rho_l = \frac{1 + |S_l|}{1 - |S_l|} = \frac{1 + |S_R| e^{-2\alpha l}}{1 - |S_R| \exp^{-2\alpha l}}. \tag{2.36}$$

2.2 Smith Chart

2.2.1 Principle of Smith Chart

To evaluate the EM-wave absorber characteristics, the Smith chart has been often utilized [4]. Hence, the principle of the Smith chart and the procedures for its treatment are described in detail here. First, let us explain the constitutive principle of the Smith chart. The Smith chart is based on the following fundamental concepts. As a rough concept of building a Smith chart, both the voltage reflection coefficient plane depicted in the polar coordinate system and the normalized impedance figure depicted in the complex plane are configured on the same plain, as shown in Figure 2.4.

The voltage reflection coefficient at the loaded end is given by the previous Eq. (2.19):

$$S_R = |S_R| \exp(j\phi_R),$$

where ϕ_R and $|S_R|$ represent the phase angle and the magnitude of the reflection coefficient, respectively. For the sake of convenience, the suffix R is omitted here.

Generally, due to the nature of the voltage reflection coefficient S in the passive circuit, the relationship $|S| \leq 1$ is held, and all reflection coefficients are represented within the circle $|S| = 1$.

In Figure 2.4a, the value of $|S|$ is depicted as a value located in any position on the concentric circle. The phase angle part φ of the reflection coefficient is represented by an angle with respect to the radius drawn from the center of the concentric circle from the phase angle $\varphi = 0$, as shown in Figure 2.4b. And the phase angle rotates counterclockwise.

Then let us try to think about the relationship between the reflection coefficient and the normalized impedance. Normalized impedance is usually represented by a complex plane, as shown in Figure 2.4d. According to the principle of mapping transformation, the complex normalized impedance plane in Figure 2.4d can be transformed, as shown in Figure 2.4e, and this complex plane can finally be depicted as in Figure 2.4f (from an intuitive perspective, it is also called a rubber transformation). Subsequently, by superimposing Figures 2.4c,f, we can construct the Smith chart with both the impedance and reflection coefficient characteristics, as shown in Figure 2.4g.

Next, let us treat the Smith chart analytically.

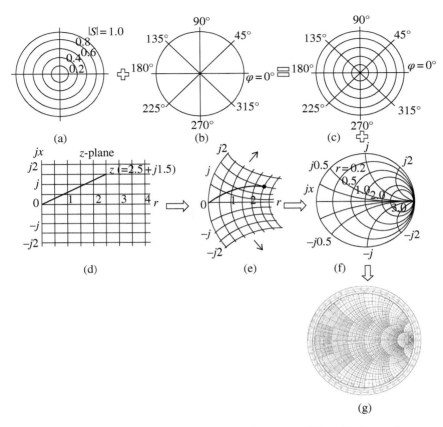

Figure 2.4 The formation of the Smith chart. (a) Reflection coefficient. (b) Phase of reflection coefficient. (c) A diagram combining (a) and (b). (d) Diagram showing complex plane. (e) Deformation process of the complex plane of (d). (f) Schematic of Smith chart. (g) Actual Smith chart. (This configuration method is called "Gum transformation" [5]. This method was proposed by Professor K. Suetake in 1952 from the Tokyo Institute of Technology on the basis of the conformal mapping concept. This is a method of hypothetical use of a rubber plate on which a figure is drawn, extending the rubber to deform the figure to finally obtain the conformal mapping figure. This method is useful for understanding the phenomenon quickly and intuitively [3].)

If representing both the reflection coefficient S and normalized impedance z using complex number expressions,

$$S = S_r + jS_i, \quad z = r + jx. \tag{2.37}$$

Since the relationship of Eq. (2.23) is satisfied between the reflection coefficient S and the normalized impedance z,

$$r + jx = \frac{1 + (S_r + jS_i)}{1 - (S_r + jS_i)}. \tag{2.38}$$

By equalizing the real part and the imaginary part of both sides of Eq. (2.38), respectively,

$$r = \frac{1 - S_r^2 - S_i^2}{(1 - S_r)^2 + S_i^2}, \quad x = \frac{2S_i}{(1 - S_r)^2 + S_i^2}. \tag{2.39}$$

After making the further arrangement of these expressions, the following expressions can be derived:

$$\left(S_r - \frac{r}{1+r}\right)^2 + S_i^2 = \left(\frac{1}{1+r}\right)^2 \tag{2.40}$$

$$(S_r - 1)^2 + \left(S_i - \frac{1}{x}\right)^2 = \left(\frac{1}{x}\right)^2. \tag{2.41}$$

Both Eqs. (2.40) and (2.41) represent the circles with coordinates (S_i, S_r).

In order to learn about the nature of these circles, if concrete numerical values are given for the coordinate components r, x, they can be drawn as illustrated in Figures 2.5a,b. As shown in Figure 2.5a, the center of the r circle moves in the positive direction of the S_r axis when the real part r representing the resistance value increases. Hence, when r goes to infinity ($r \rightarrow \infty$), the circle that depends on r converges to one point $S_r = 1.0$ on the S_r axis.

In addition, as shown in Figure 2.5b, the circle centers of the imaginary part x, which represents the reactance values, are located on the $S_r = 1.0$ line in response to a positive or a negative value of x. In the Smith chart, the normalized input impedance z is expressed only by the part within the unit circle from the condition that the absolute value of the reflection coefficient in the usual

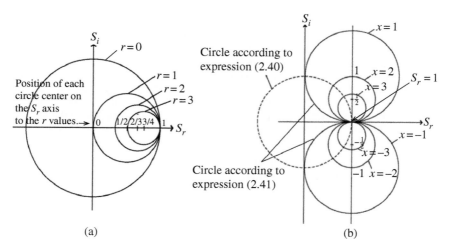

(a)   (b)

Figure 2.5 Construction process of the Smith chart. (a) Circle group by Eq. (2.40) and (b) circle group by Eq. (2.41).

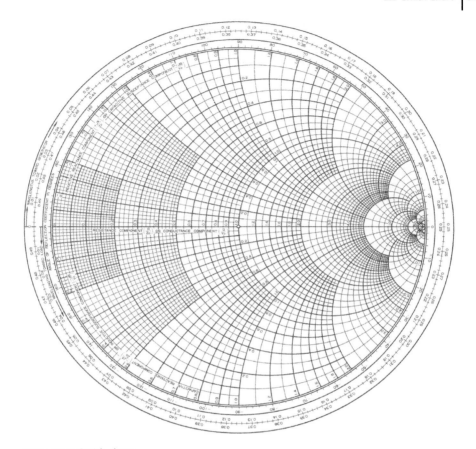

Figure 2.6 Smith chart.

passive circuit is less than 1.0. Accordingly, by integrating Figures 2.5a,b into one drawing, the Smith chart can be configured. Figure 2.6 shows the actual Smith chart.

Summarizing the main points of the Smith chart before describing their details:

(a) In the Smith chart, to avoid complication, the polar coordinate scale of the reflection coefficient is not depicted.

(b) The normalized reactance x can take positive and negative signs, the positive normalized reactance is in the upper half circle, and the negative normalized reactance is in the lower half circle. The locus of the point at $x = 0$ is located on the S_r axis.

(c) The absolute value of the reflection coefficient $|S|$ is 0 at the center point of the Smith chart and normalized resistance r and reactance x takes values of

$r = 1$ and $x = 0$, respectively. Then, the value of the voltage standing-wave ratio in this case is 1.

(d) On the positive S_r axis of the Smith chart, it corresponds to the point where the phase angle of the reflection coefficient is 0 and the voltage value of the standing wave becomes the maximum. Also, on the negative S_r axis, it corresponds to the point where the phase angle of the reflection coefficient φ takes π and the voltage value of the standing wave becomes the minimum.

(e) On the outer periphery of the Smith chart, the value of wave number l/λ indicating the generator direction and the load direction is depicted.

Let us complement these contents further.

As for this reflection coefficient in the Smith chart, it is suggested that the degree of reflection can be expressed by the standing-wave ratio ρ from the relationship of Eq. (2.25). The standing-wave ratio actually scales in the range $r \geq 1.0$ of S_r on the real axis of the Smith chart. Then it becomes possible to calculate the $|S|$ from the standing-wave ratio ρ which is designated in the Smith chart. Let us touch briefly on this point.

It is apparent from the following expression that ρ becomes equal to $r(r \geq 1.0)$ in this manner when the imaginary part of the normalized impedance is 0. Namely, since $x = 0$ in Eq. (2.38), then $S_i = 0$, and the following relation holds:

$$S_r = \frac{r-1}{r+1}, \tag{2.42}$$

where $r \geq 1.0$.

Here, with reference to Eq. (2.25),

$$\rho = \frac{1 + |S_r|}{1 - |S_r|}. \tag{2.43}$$

Substituting Eq. (2.42) into Eq. (2.43), the following relationship holds:

$$\rho = \frac{1 + (r-1)/(r+1)}{1 - (r-1)/(r+1)} = r. \tag{2.44}$$

This is the reason why the resistance value r on the S_r axis in the Smith chart can be connected with the standing-wave ratio ρ.

Next, on the Smith chart, let us think about how to represent the phase term of the reflection coefficient. The phase angle φ of the reflection coefficient, as described previously, is represented using an angle from the base line of the S_r axis. By the way, the normalized impedance in the Smith chart is often treated in relation to the normalized impedance on the TL in many cases. In order to enable such treatment, we have to associate the distance l from the load of the TL with the phase angle of the reflection coefficient.

Let us consider the relationship between the distance l from the load end on the TL and the phase angle of the reflection coefficient. By the way, the phase angle of the reflection coefficient and the moving distance from the load end are related to the normalized impedance as follows. First, using Eq. (2.10), we can derive the normalized impedance z_l at the distance l that looks toward the load end on the TL.

$$V_l = V_i e^{j\beta l}[1 + S_R e^{-j2\beta l}] \tag{2.45}$$

$$I_l R_c = V_i e^{j\beta l}[1 - S_R e^{-j2\beta l}]. \tag{2.46}$$

From these equations, the normalized input impedance z_l at the distance l on the TL is

$$z_l = \frac{1 + S_R e^{-j2\beta l}}{1 - S_R e^{-j2\beta l}}. \tag{2.47}$$

By substituting Eq. (2.19) into Eq. (2.47),

$$z_l = \frac{1 + |S_R|e^{j(\varphi_R - 2\beta l)}}{1 - |S_R|e^{j(\varphi_R - 2\beta l)}} = \frac{1 + |S_R|e^{j\varphi_l}}{1 - |S_R|e^{j\varphi_l}}, \tag{2.48}$$

where $\varphi_l = \varphi_R - 2\beta l$.

Equation (2.48) represents the relationship between the normalized impedance z_l facing the load end at any point l on the TL and the reflection coefficient S_R at the load end. In order to investigate a reference point at l in the Smith chart, let us here think about the case where the load impedance R_R will be considered as pure resistance and will be less than the characteristic impedance R_c of the TLs.

That is, since the relationship $Z_R = R_R < R_C$ is preserved, the normalized impedance at the load impedance z_R can be expressed in the form of $z_R < 1.0$. Thus, the reflection coefficient at the load position becomes $S_R < 0$ from Eq. (2.19). Here, from the conditions $\cos\varphi_R = -1$ and $j\sin\varphi_R = 0$, the condition that the reflection coefficient S_R at the load end of expression (2.19) becomes negative. That is, $\varphi_R = 180°$.

Because l is zero at the load end, it can be found from Eq. (2.48) that the relation $\varphi_l = \varphi_R = 180°$ is held. Further, the reflection coefficient at the load end $l = 0$ exists on the real axis between $0 < S_r < 1$ under the condition $z_R < 1.0$. Consequently, this line is defined as the origin of the phase angle.

Next, consider the direction of rotation of the phase angle in the present case. Regarding the rotation direction of the phase angle, the following relations exist. As is apparent from Eq. (2.48), the phase angle φ_l decreases when the consideration point l on the TL moves toward the generator direction. Therefore, this relation indicates the reverse direction with respect to the direction of $\varphi(= \varphi_R)$ which was initially defined in Figure 2.4c.

This is the reason why the generator direction or the load direction is printed on the outer circumference of the Smith chart from reference axis on $-1 < S_r < 0$ (Impedance is 0), and is related to the item (e).

Moreover, in practice, the Smith chart is not scaled by the phase angle $2\beta l$. Since it is inconvenient to calculate this angle in the actual Smith chart, the value of the wave number l/λ instead of the phase angle $2\beta l$ is scaled. That is, in the Smith chart, attention is focused on l in $2\beta l$. The relationship between this phase angle $2\beta l$ and the wave number l/λ is expressed by the following equation:

$$2\beta l = 2(2\pi/\lambda)l = 4\pi(l/\lambda)(= 720°(l/\lambda)).$$

2.2.2 Admittance Chart

We often encounter problems when dealing with the case where the circuit elements are connected in parallel on the TL. In such a case, it is convenient to consider the Smith chart as an admittance representation. Now, the normalized admittance, facing toward the load from the distance l on the TL, can be expressed by the following equation after taking the reciprocal of Eq. (2.15).

$$y_l = \frac{1}{z_l} = \frac{\cos \beta l + j z_R \sin \beta l}{z_R \cos \beta l + j \sin \beta l}$$
$$= \frac{z_R \sin \beta l - j \cos \beta l}{\sin \beta l - j z_R \cos \beta l}. \tag{2.49}$$

On the other hand, in the normalized impedance expression, when the value of the length l is changed to $l + \lambda/4$ on the TL, in other words, when the phase angle is rotated by $180°$, the normalized impedance can be represented by the following equation:

$$z_l|_{l=x+\lambda/4} = \frac{z_R \cos \beta(l + \lambda/4) + j \sin \beta(l + \lambda/4)}{\cos \beta(l + \lambda/4) + j z_R \sin \beta(l + \lambda/4)}$$
$$= \frac{z_R \sin \beta l - j \cos \beta l}{\sin \beta l - j z_R \cos \beta l} \equiv y_l. \tag{2.50}$$

As is apparent from this equation, in the case where the phase angle is rotated by $180°$, the value on the Smith chart represents the value of admittance.

This fact means that the admittance-type Smith chart can be designated by rotating by $180°$, as shown in Figure 2.7. Therefore, in the values on the admittance chart, the resistance value r directly represents the conductance g. Also, the imaginary part x of the normalized impedance value z represents the susceptance b, and the $-x$ can be expressed as the susceptance $-b$.

Concerning the standing-wave ratio, we can use the conductance value of g on the negative axis of S_r in the region $b = 0$, $g > 1$. In the same way as the impedance chart, the upper semicircular part on the admittance chart represents the inductive characteristic region, and the lower semicircular part represents a capacitive characteristic region. However, the scale based on the wave number l/λ, which is depicted on the peripheral portion, cannot be utilized

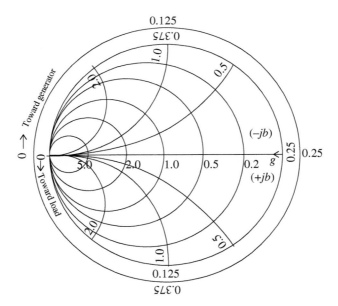

Figure 2.7 Admittance chart.

in the same way as in the case of the impedance chart. This is because on the Smith chart rotated by 180°, the scale of the point $b = 0$ on the S_r axis, which should become a reference to the wave number l/λ, is expressed as 0.25.

Therefore, this is necessary to convert the reference point value to zero when using the admittance chart. That is, this is because the reference of the phase angle on the TL does not change even if considering admittance. In addition, as shown in the admittance diagram in Figure 2.7, arrows indicating "Wavelengths toward generator" and "Wavelengths toward load" can be used as they are.

2.2.3 Examples of Smith Chart Application

In the study of the EM-wave absorber characteristics (matching characteristics), the Smith chart has been often utilized. In this section, let us introduce a few specific examples of how to use the Smith chart.

2.2.3.1 Impedance of Transmission Line with Short-circuit Termination

First, as a fundamental matter of treatment of the Smith chart, the relationship between the position of the TL and the normalized impedance is described. In this case, the load at the end of the TL is short-circuited. Also, the reflection coefficient is $S_l = 1$, the standing-wave ratio is $\rho = \infty$, and the phase angle is $\varphi_l = 180°$.

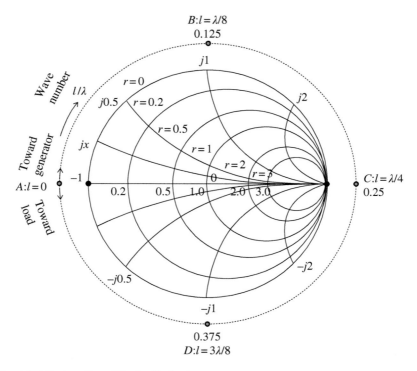

Figure 2.8 Explanation of the Smith chart.

We consider here the problem of the normalized input impedance facing the load termination at a point which is moved by l from the load to the EM-wave generator direction. The normalized load impedance in this case can be expressed as $z_i = 0 + 0j$ with $l = 0$. Accordingly, this impedance is represented by a point A on the Smith chart, as shown in Figure 2.8. The impedance at a certain position l on the TL in the direction of the transmitter from the load is given by the wave number l/λ in the oscillator direction in the Smith chart as described previously.

For example, when l takes the value of $l = \lambda/8$, the point B in Figure 2.8, that is routed by 90° with respect to A, is a new moving point.

Likewise, if l is $\lambda/4$, the point is given by the point C, which is rotated by 180° from A, and if l is $3\lambda/8$, the point is depicted by D which is rotated by 270°, as shown in Figure 2.8. Each of the impedances is given by $z_l = 0 + j1$, $z_l = 0 + j\infty$, and $z_l = 0 - j1$, respectively.

2.2.3.2 Matching Method with a Single Movable Stub
As one method of taking a matching on the TL, there is a method of connecting a stub in the middle of the TL.

In this section, as an example of an application of the Smith chart, let us consider the method of matching using stubs. Deepening the knowledge of the matching method on the TL using the Smith chart becomes useful when making the matching for the EM-wave absorber which can be represented by replacing it with an equivalent TL.

First, let us consider an example of a matching problem in a normal TL terminated by the load Z_R shown in Figure 2.9. This matching method can be achieved by connecting a single stub with a short circuit in the middle of the main TL terminated with load $y_R(= g + jb)$ and adjusting the length of each TL, l_1 and l_2, as shown in Figure 2.10.

These procedures are summarized as follows:

(a) First, the length l_1 is adjusted on the main TL on which the standing wave occurs, so as to take the value of $y_a = 1 + jb$. A point satisfying this condition is denoted by a.

(b) When connecting stubs in parallel with the main TL, the normalized admittance value of the stub is adjusted so as to take the value of $y_s = -jb$ by adjusting the stub length l_2. As a result, the total normalized admittance looking from the power supply side b toward the load terminal becomes $y_t = (1 + jb) - jb = 1$, which means that matching has been achieved.

Although the way of taking the matching in this case can be calculated theoretically based on these methods, this matching can also be easily determined using the Smith chart.

Figure 2.9 Transmission model terminated by a load with Z_R.

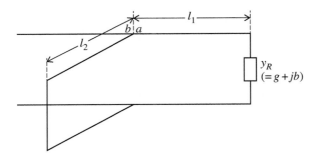

Figure 2.10 Explanation of a single-stub matching method.

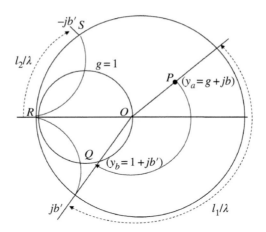

Figure 2.11 Explanation on how to treat the Smith chart.

Now, as shown in Figure 2.11, let us deal here with the admittance chart in the Smith chart.

(a) When representing the load impedance Z_R in the form of normalized admittance $y_a = g + jb$, this admittance value is now assumed to be denoted by a point P in Figure 2.11.

(b) As in the case of the theoretical treatment described, we have to determine the distance l_1 from the load, that is, a point a where the input admittance takes the value of $y_a = 1 + jb$. This becomes possible by drawing a circle toward the generator side with OP as the radius from the center O of the admittance chart and finding the intersection point Q with the circle $g = 1$. By this manipulation, the susceptance value jb changes to jb' at point Q, and the conductance takes the value of 1.0.

Hence, the admittance value is expressed as $y_b = 1 + jb'$ at point Q on the Smith chart.

(c) Next, the length of l_2 in the stub shown in Figure 2.10 has to be determined, but since the susceptance value corresponding to a point Q is jb', the susceptance value which needs to cancel jb' is provided by the value of a point S positioned symmetrically against the value jb'. Therefore, the line length l_2 of the stub when taking the value of $-jb'$ is obtained from the wave number l_2/λ between R and S. In this way, using the Smith chart, it becomes possible to take the matching of this TL.

2.2.3.3 Matching Method Using Fixed Multiple Stubs

Next, let us consider the matching method of the TL when using two stubs, as shown in Figure 2.12, from the perspective of the Smith chart. In this case, the positions *a* and *c* of the two stubs are fixed on the TL, unlike in the previous single stub. Thus, admittance values can be adjusted by changing the stub's lengths l_1 and l_2.

(a) At the point *a* on the main TL, let $y_a = g + jb$ be the normalized admittance in the case of looking at the load terminal with Z_R. It is assumed that the value of y_a at this time is plotted at point *P* in the Smith chart in Figure 2.13.

(b) Next, by changing the length l_1 of the stub S_1, the susceptance $y_{s1} = jb_{s1}$ of the stub is added to y_a. This normalized admittance $y_b (= g + jb + y_{s1})$ means the normalized admittance which looks at the load side from point *b* on the main TL including the stub S_1.

As shown in Figure 2.13, note here that by adjusting l_1, the value of y_{s1} is set so that point *P* takes the value of point *Q* on the dotted line circle, which is depicted by inverting the circle $g = 1$ by 180°.

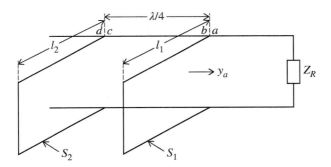

Figure 2.12 The case of transmission line with multiple stubs.

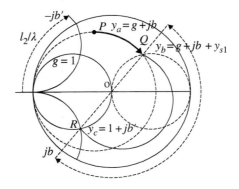

Figure 2.13 Matching method explanation in the case of double stubs.

In this operation, the point P must be moved to the point Q without changing the conductance value in the conductance circle by keeping its value constant. That is, only the susceptance value has to be changed.

The intention to use the dotted circle in Figure 2.13 is in that the real part of the total normalized admittance when looking at the load side from point d can take the value of 1.0.

(c) To see this, y_b at point c which is apart from point b by $\lambda/4$ on the TL is equivalent to finding the value of admittance y_c of the R point. This value y_c can be obtained by rotating the point Q (y_b value) in the generator direction by 180°. Since this point R is on the circumference of $g = 1$, it takes the form $y_c = 1 + jb'$.

Next, it is necessary to adjust the stub length l_2 in the stub S_2 so that $y_{s2} = -jb'$ can be added to y_c. Since the susceptance value at the point R is jb', the susceptance value of $-jb'$ is given by the susceptance value with respect to the point symmetrical to this point R. Therefore, the line length l_2 of the stub S_2 is presented from the wave number l_2/λ when taking the value of $-jb'$, as shown in Figure 2.12.

Consequently, the total normalized admittance y_t that looks at the load terminal from point d can be set to 1.0, and this TL having a load can be matched.

2.3 Fundamentals of Electromagnetic Wave Analysis

In the previous section, the basic matters for understanding the wave absorber characteristics were explained from the standpoint of the TL theory.

Here, let us first briefly explain the methods of deriving Maxwell's equations, which play an important role in the analysis of EM-wave absorbers, and then the derivation of wave equations. Using the EM fields derived from these equations, the problem of reflection at the normal incidence on a perfect conductor is first investigated. Further, the reflection and transmission coefficients in the case of normal and oblique incidence, which consist of the structure of two media with different medium constants, are explained. Finally, the multiple reflection problems as to a medium of thickness d having a predetermined electric constant, which is needed as a fundamental knowledge of the EM-wave absorber design, are briefly introduced.

2.3.1 Derivation of Maxwell's Equations

Maxwell's equations can be derived as the first EM equation based on Ampere's circle integral law, and as the second EM equation based on Faraday's EM induction law, respectively.

2.3.1.1 Maxwell's First Electromagnetic Equation

The Maxwell equation, called the first EM equation based on Ampere's law of circle integration, can be derived in the following manner. Let us consider a minute rectangle $ABCD$ on the yz plane as shown in Figure 2.14, and assume the current which is interlinking perpendicular to the rectangular circuit in the x-axis direction.

To apply the following Ampere's circuital law, we first set the magnetic fields along each side in the rectangular circuit:

$$\oint \boldsymbol{H}\cdot d\boldsymbol{l} = I. \tag{2.51}$$

Here, the electric field, magnetic field, and current density of the point $P\ (x, y, z)$ are denoted as $\boldsymbol{E}(E_x, E_y, E_z)$, $\boldsymbol{H}(H_x, H_y, H_z)$, $\boldsymbol{i}(i_x, i_y, i_z)$, respectively. The sides AB and DC (corresponding to the integrating paths) are separated by $\pm \Delta z/2$ on the z axis from the point P, respectively.

Therefore, the magnetic fields H_{yAB}, H_{yDC} on the AB and DC sidelines vary slightly with respect to the magnetic field H_y which is located at the point P.

Assuming now that the amount of change in the magnetic field in a section Δz is ΔH_y, the rate of change between Δz is $H_y/\Delta z$; so the magnetic field H_{yAB}, H_{yDC} on the AB or DC sidelines can be expressed using these incremental Δz values as follows:

$$H_{yAB}, H_{yDC} = \big(\text{Magnetic field } H_y \text{ at point P} + \text{Change amount of magnetic}$$
$$\text{field due to shift by} \pm \Delta z/2 \text{ to AB or DC sideline}\big).$$

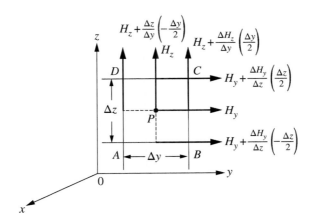

Figure 2.14 The positional relationship of the magnetic field in a small rectangle *ABCD* in the *yz* plane.

Therefore,

$$H_{yAB} = H_y + \left\{ \frac{\Delta H_y}{\Delta z} \left(-\frac{\Delta z}{2} \right) \right\} \tag{2.52}$$

$$H_{yDC} = H_y + \left\{ \frac{\Delta H_y}{\Delta z} \left(\frac{\Delta z}{2} \right) \right\}. \tag{2.53}$$

Similarly, the magnetic field along the sidelines *BC* and *AD* is also,

$$H_{zBC} = H_z + \left\{ \frac{\Delta H_z}{\Delta z} \left(\frac{\Delta y}{2} \right) \right\} \tag{2.54}$$

$$H_{zAD} = H_z + \left\{ \frac{\Delta H_z}{\Delta y} \left(-\frac{\Delta y}{2} \right) \right\}. \tag{2.55}$$

Next, let us consider the current equivalent to the right side of Eq. (2.51). When the current density is i_x [A/m^2], letting Δi_x be the total current flowing in the x direction within the infinitesimal area $\Delta y \Delta z$ as shown in Figure 2.15,

$$\Delta i_x = i_x (\Delta y \Delta z). \tag{2.56}$$

In these relations, in order to express the Ampere's circuit law of Equation (2.51), we multiply each of the magnetic field equations (2.52)–(2.55) along the rectangular circuit by the corresponding length Δy or Δz, respectively, and then put the summation of these expressions so as to be equal to (2.56).

That is,

$$H_{yAB}\Delta y - H_{yDC}\Delta y + H_{zBC}\Delta z - H_{zAD}\Delta z = i_x(\Delta y \Delta z). \tag{2.57}$$

Substituting the expressions of Eqs. (2.52)–(2.55) to the left side of Eq. (2.57),

$$\left(\frac{\Delta H_z}{\Delta y} - \frac{\Delta H_y}{\Delta z} \right) \Delta y \Delta z = i_x (\Delta y \Delta z). \tag{2.58}$$

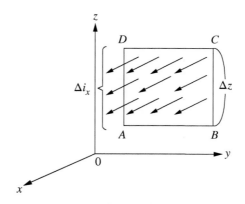

Figure 2.15 Current interlinked to a small rectangle *ABCD*.

Here, let i_x be the x component of the total current density. i_x can be expressed by the following equation, since it consists of the sum of the conduction current per unit area and the displacement current:

$$i_x = \kappa E_x + j\omega\varepsilon E_x \ [\text{A/m}^2], \tag{2.59}$$

where κ is the conductivity, ε the permittivity of medium, and ω the angular frequency.

By substituting Eq. (2.59) into Eq. (2.58) and considering the limit where $\Delta x, \Delta y, \Delta z$ are close to zero,

$$\frac{\partial H_z}{\partial y} - \frac{\partial H_y}{\partial z} = \kappa E_x + j\omega\varepsilon E_x. \tag{2.60}$$

In exactly the same way, considering the $x - y$ and $z - x$ planes, the following equations can be derived:

$$\frac{\partial H_x}{\partial z} - \frac{\partial H_z}{\partial x} = \kappa E_y + j\omega\varepsilon E_x \tag{2.61}$$

$$\frac{\partial H_y}{\partial x} - \frac{\partial H_x}{\partial y} = \kappa E_z + j\omega\varepsilon E_z. \tag{2.62}$$

Rewriting these equations in vector form,

$$\nabla \times H = \kappa E + j\omega\varepsilon E = (\kappa + j\omega\varepsilon)E \ \ [\text{A/m}^2]. \tag{2.63}$$

Equation (2.63) represents the Maxwell's first EM equation.

Generally, in addition to the conduction current of the first term on the right-hand side of Eq. (2.63) and the displacement current of the second term, the convection current and the induced current exist as currents. However, in most cases, wave phenomena in a medium, such as an EM-wave absorber, can be analyzed using these conduction and displacement currents.

2.3.1.2 Maxwell's Second Electromagnetic Equation

Next, let us derive Maxwell's second EM equation based on Faraday's EM induction law. The integral representation of Faraday's EM induction law is

$$\oint E \cdot dl = -\frac{d\Phi}{dt}. \tag{2.64}$$

As in Section 2.3.1.1, consider a small rectangular circuit $ABCD$ in the yz plane, as shown in Figure 2.16. In this case, the electric field at point P is represented by E_y, E_z.

In this $ABCD$ circuit, it is found that the induced voltage in the left side of Eq. (2.64) is calculated by multiplying the electric field component of each side by the length of each side. The meaning of Eq. (2.64) is that the sum of these induced voltages is equal to the time change of the total magnetic flux passing through this rectangular circuit with a negative sign.

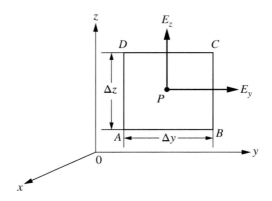

Figure 2.16 The positional relationship of the electric field in a small rectangle *ABCD* in the *yz* plane.

First, similarly to Eq. (2.57) in Section 2.3.1.1, since the electric field $E_{yAB}, E_{yDC}, E_{zBC}, E_{zAD}$ along each side can be expressed in terms of the electric field components E_y, E_z, the products of these electric fields by the corresponding side length are given by the following induced voltage expression:

$$\left(\frac{\Delta E_z}{\Delta y} - \frac{\Delta E_y}{\Delta z} \right) \Delta y \Delta z. \tag{2.65}$$

Next, let us consider the right side of Eq. (2.64). When the *x* component of the magnetic flux density interlinked to this rectangular circuit *ABCD* is B_x as shown in Figure 2.17, this can be expressed as Figure 2.17

$$\Delta B_x = \mu H_x (\Delta y \Delta z).$$

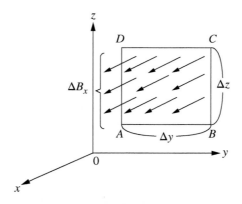

Figure 2.17 Magnetic flux interlinking perpendicularly to a small plane *ABCD*.

Since this x component of magnetic flux density varies with time at the angular frequency ω, the time change rate of B_x is

$$\frac{\partial}{\partial t}\mu H_x(\Delta y \Delta z) = j\omega\mu H_x(\Delta x \Delta z). \tag{2.66}$$

Here, according to Lenz's law which regulates the direction of the EM induction regarding Faraday's law, it is necessary to make the expression (2.66) negative when equalizing Eqs. (2.65) and (2.66).

Hence,

$$\frac{\Delta E_z}{\Delta y} - \frac{\Delta E_y}{\Delta z} = -j\omega\mu H_x. \tag{2.67}$$

When the rectangular circuit $ABCD$ is minimized to the infinitely small, Δ should be expressed as $\Delta \rightarrow \partial$,

$$\frac{\partial E_z}{\partial y} - \frac{\partial E_y}{\partial z} = -j\omega\mu H_x. \tag{2.68}$$

Similarly, for the $z - x$ and $x - y$ planes,

$$\frac{\partial E_x}{\partial z} - \frac{\partial E_z}{\partial x} = -j\omega\mu H_y \tag{2.69}$$

$$\frac{\partial E_y}{\partial x} - \frac{\partial E_x}{\partial y} = -j\omega\mu H_z. \tag{2.70}$$

By unifying these scalar equations with the vector representation, Maxwell's second EM equation can be derived:

$$\nabla \times E = -j\omega\mu H \ [\text{V/m}^2]. \tag{2.71}$$

These first and second EM equations are collectively referred to as Maxwell's EM equations. In addition, since they are presented in a differential form, they can express a relationship satisfied at an arbitrary point in space.

2.3.2 Wave Equations

Before describing the wave equations, let us briefly introduce the general expression of Maxwell's equations. And then, the method of derivation of the wave equations is introduced.

As for the Maxwell's equations described in the previous section, they can be expressed as the generalized symmetrical type of Maxwell's equations. These generalized types of Maxwell's equations can be expressed by the following expressions:

$$\nabla \times H = \frac{\partial D}{\partial t} + J \tag{2.72}$$

$$\nabla \times E = -\frac{\partial B}{\partial t} - M. \tag{2.73}$$

In general, the current includes the conduction current, the displacement current, the convection current, the polarization current, and so on. But in the discussion here, we simply consider the electric current density on the basis of free electrons flowing in proportion to the applied electric field in the conductive medium. $\partial D/\partial t$ represents the so-called displacement current density, which flows even in vacuum or a dielectric medium. In addition, M in Eq. (2.72) is the fictitious magnetic current density, and $\partial D/\partial t$ is the displacement magnetic flux density (see Appendix 2.A.1).

This concept of "magnetic current" is applied to theoretical analysis in antenna application fields, such as a slot antenna. Also, the generalized Gauss' laws for electric and magnetic fields are given by the following equations, where ρ_e is the electric charge density and ρ_m is the fictitious magnetic charge density:

$$\nabla \cdot D = \rho_e \tag{2.74}$$

$$\nabla \cdot B = \rho_m. \tag{2.75}$$

Now, since here we consider a homogeneous isotropic medium without existing electrons, ions, etc., only the conduction current and the displacement current should be taken into account in Eq. (2.72). Using material constants, permittivity ε, permeability μ, conductivity κ, and volume charge density ρ, Maxwell's equations in the present case are expressed as follows:

$$\nabla \times E = -\mu \frac{\partial H}{\partial t} \tag{2.76}$$

$$\nabla \times H = \varepsilon \frac{\partial E}{\partial t} + \kappa E. \tag{2.77}$$

Further, since we consider here a medium in which there is no electric charge and magnetic charge in the present medium, the right-hand sides in Eqs. (2.74) and (2.75) are treated as 0. In these equations, when the electric and magnetic fields vary with the angular frequency $\omega = 2\pi f$, they are expressed by the following equation, where f is the frequency:

$$E = E_0 e^{j\omega t} \tag{2.78}$$

$$H = H_0 e^{j\omega t}. \tag{2.79}$$

Using the relations of $\partial E/\partial t = j\omega E$ and $\partial H/\partial t = j\omega H$, Eqs. (2.76) and (2.77) can be represented as

$$\nabla \times E = -j\omega \mu H \tag{2.80a}$$

$$\nabla \times H = (\kappa + j\omega \varepsilon)E. \tag{2.80b}$$

Here, if the magnetic field on the right-hand side of Eq. (2.80a) is substituted into (2.80b) to eliminate the magnetic field H,

$$\nabla \times \nabla \times E = -j\omega \mu (\kappa + j\omega \varepsilon)E.$$

Applying the following equations to the given relation,

$$\left.\begin{array}{l} \nabla \times \nabla \times \mathbf{A} = \nabla\nabla\cdot\mathbf{A} - \nabla^2\mathbf{A} \\ \nabla\cdot\mathbf{A} = \nabla\cdot\mathbf{E} = \rho = 0 \end{array}\right\}.$$

Hence, the following expression is given:

$$\nabla^2 E + \gamma^2 E = 0, \tag{2.81}$$

where $\gamma^2 = j\omega\mu(\kappa + j\omega\varepsilon)$.

Similarly, the following wave equation for the magnetic field yields

$$\nabla^2 H + \gamma^2 H = 0. \tag{2.82}$$

As an example, in Eq. (2.81), when the wave equation for the electric field is expressed by x, y, z coordinates and only the x component is considered, it can be expressed by the following simple equation (see Appendix 2.A):

$$\nabla^2 = \frac{\partial}{\partial x^2} + \frac{\partial}{\partial y^2} + \frac{\partial}{\partial z^2}, \gamma^2 = j\omega\mu(\kappa + j\omega\varepsilon)$$

$$\mathbf{E} = i E_x + j E_y + k E_z \quad \mathbf{H} = i H_x + j H_y + k H_z \text{ (In rectangular coordinate)}$$

$$\frac{\partial^2 E_x}{\partial x^2} + \frac{\partial^2 E_x}{\partial y^2} + \frac{\partial^2 E_x}{\partial z^2} + \gamma^2 E_x = 0. \tag{2.83}$$

2.3.3 Reflection from Perfect Conductor in Normal Incidence

As a basic problem in the wave absorber theory, the problem of reflection from a perfect conductor plane is first taken up here. Generally, the electromagnetic wave power incident on the boundary between two media is represented by reflected power and transmitted power as shown in Figure 2.18.

When considering the general case, where EM waves are obliquely incident on the perfect conductor plane, as shown in Figure 2.18, a plane including both the direction of the incident power $\mathbf{P}_i$ and the normal vector $\mathbf{n}$ on the boundary surface is called the "plane of the incidence."

The angles $\theta_i, \theta_r, \theta_t$ between $\mathbf{n}$ and $\mathbf{P}_i$, $\mathbf{n}$ and $\mathbf{P}_r$, and $\mathbf{n}$ and $\mathbf{P}_t$ are called the angle of incidence, the angle of reflection, and the angle of transmission, respectively. In this section, first of all, the normal incident case ($\theta_i = 0$) of the transverse electric (TE) wave on a perfect conductor, as shown in Figure 2.19, is investigated.

As shown in Figure 2.19, the equation of the incident wave propagating along the $+z$ direction with the amplitude A and phase constant γ_1 is given by

$$E_{iy}(z) = A e^{-j\gamma_1 z}. \tag{2.84}$$

Here, $\gamma_1 = \omega\sqrt{\varepsilon_1 \mu_1}$.

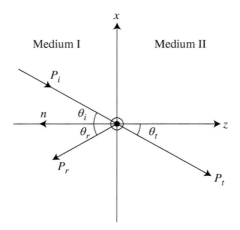

Figure 2.18 Explanation of the oblique incidence into general mediums 1 and 2.

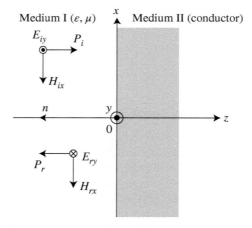

Figure 2.19 Explanation of the normal incidence.

Between the magnetic field and the electric field, in this case the relationship $H = -j1/\omega\beta\mu_1 \cdot \partial E_{iy}/\partial z$ is established from Maxwell's equations, as follows:

$$H_{ix} = -\frac{k_1}{\omega\mu_1}Ae^{-j\gamma_1 z}. \qquad (2.85)$$

Also, the reflected wave of the electric field can be expressed in terms of B instead of the amplitude A in Eq. (2.84), and the sign of the exponential phase term changes as

$$E_{ry} = Be^{j\gamma_1 z}. \qquad (2.86)$$

In the same manner as in expression (2.85), the reflected magnetic field is

$$H_{rx} = \frac{k_1}{\omega\mu_1} Be^{j\gamma_1 z}.$$

In medium I, since the incident wave and the reflected wave exist simultaneously, the total electric field in region 1 is

$$E_{y1}(z) = E_{iy}(z) + E_{ry}(z) = Ae^{-j\gamma_1 z} + Be^{j\gamma_1 z}. \tag{2.87}$$

By applying the boundary condition on the conductor surface in this equation, the amplitude relationship between A and B can be determined.

Since the "tangential component of the electric field is zero" on the perfect conductor surface, when this boundary condition is applied to Eq. (2.87),

$$E_{y1}(z)|_{Z=0} = Ae^{-j\gamma_1 z} + Be^{j\gamma_1 z}|_{z=0}. \tag{2.88}$$

Hence,

$$A = -B. \tag{2.89}$$

The absolute values of both the incident electric field and the reflected field are equal, but their positive and negative signs are opposite.

Next, substituting Eq. (2.89) into Eq. (2.87) and applying the formula $e^{\pm jx} = \cos x \pm j \sin x$, the following expression can be derived:

$$E_{y1}(z) = A(e^{-j\gamma_1 z} - e^{j\gamma_1 z}) = -2jA \sin \gamma_1 z. \tag{2.90}$$

On the other hand, in terms of the relation of (2.89), the following expression for the magnetic field can be derived in the same way as for the electric field:

$$H_{x1}(z) = H_{ix}(z) + H_{rx}(z) = -\frac{\gamma_1}{\omega\mu_1}(Ae^{-j\gamma_1 z} - Be^{j\gamma_1 z})$$

$$= -\frac{\gamma_1 A_2}{\omega\mu_1}\left(\frac{e^{-j\gamma_1 z} + e^{j\gamma_1 z}}{2}\right). \tag{2.91}$$

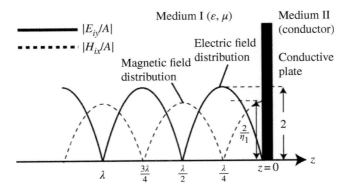

Figure 2.20 Standing wave distributions.

Hence,

$$H_{x1}(z) = -\frac{2A}{\eta_1}\cos\gamma_1 z, \tag{2.92}$$

where $\eta_1 = \sqrt{\mu_1/\varepsilon_1}$, η_1 is the electromagnetic wave characteristic impedance (intrinsic impedance) in medium I.

As estimated from Eqs. (2.90) and (2.92), the incident electric field wave, and the magnetic field wave and their reflected waves are synthesized, in medium I.

As a result, the standing waves arise, as shown in Figure 2.20.

2.3.4 Reflection and Transmission in Two Medium Interfaces

2.3.4.1 Normal Incidence Cases

To further deepen the understanding of the treatment methods of the reflection coefficient, let us investigate here the reflection problems on a boundary surface composed of two media extending to infinity as shown in Figure 2.21. The values of the permittivity and permeability in medium I and II are denoted by (ε_1, μ_1) and (ε_2, μ_2), respectively.

First of all, let us consider the reflection and transmission characteristics when a plane TE wave propagates in the z-axis direction with the normal incidence on the boundary surface (Figure 2.21).

Assuming that the electric fields have only the vertical component E_y in the incident plane, the electric field components x and y vanish.

This case is usually called a one-dimensional problem. Hence, the conditions $\partial/\partial x = 0$ and $\partial/\partial y = 0$ are imposed in the wave equation (2.83) in Section 2.3.2. Also, the propagation constants of mediums I and II are denoted by γ_1 and γ_2, respectively.

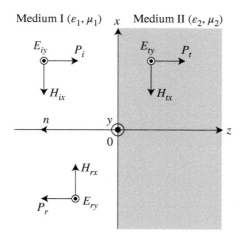

Figure 2.21 Normal incidence case at different medium interfaces.

Accordingly, the equations in the medium I can be represented as follows:

Medium I:

$$\left.\begin{array}{l} \dfrac{\partial^2 E_{y1}}{\partial z^2} + \gamma_1^2 E_{y1} = 0 \\[2mm] H_{x1} = -j\dfrac{1}{\omega \mu_1}\dfrac{\partial E_{y1}}{\partial z} \end{array}\right\}, \tag{2.93}$$

where $\gamma_1 = \omega\sqrt{\varepsilon_1 \mu_1}$. The solution of Eq. (2.93) can be expressed as the sum of the incident wave and the reflected wave in the form of Eq. (2.87) in Section 2.3.3.

$$E_{y1}(z) = E_{iy}(z) + E_{ry}(z)$$
$$= Ae^{-j\gamma_1 z} + Be^{j\gamma_1 z}$$

(Incident wave)(Reflected wave) $(z < 0)$.   (2.94)

Similarly, as for the magnetic field,

$$H_{x1}(z) = H_{ix}(z) + H_{rx}(z)$$
$$= -(1/\eta_1)(Ae^{-j\gamma_1 z} - Be^{j\gamma_1 z}), \tag{2.95}$$

where $\eta_1 = \sqrt{\mu_1/\varepsilon_1}$. As for the EM fields in the medium II, they are presented by the following equations:

Medium II:

$$\dfrac{\partial^2 E_{y2}}{\partial z^2} + \gamma_2^2 E_{y2} = 0$$
$$H_{x2} = -j\dfrac{1}{\omega \mu_2}\dfrac{\partial E_{y2}}{\partial z}. \tag{2.96}$$

The general solution of the electric field in Eq. (2.96),

$$E_{y2}(z) = Ce^{-j\gamma_2 z} + De^{j\gamma_2 z}.$$

(Electric field in medium II)(Traveling wave)(Backward wave)   (2.97)

In this case, since the medium II is extended infinitely in the z-axis direction, the reflected wave does not exist. Therefore, $D = 0$ in Eq. (2.97). Then, the expression (2.97) is

$$E_{y2} = E_{ty} = Ce^{-j\gamma_2 z}$$

(Electric field in medium II) $(z > 0)$ (2.98)

and

$$H_{x2} = H_{tx} = -(C/\eta_2)e^{-j\gamma_2 z}.$$

(Magnetic field in medium II)   (2.99)

Here, $\eta_2 = \sqrt{\mu_2/\varepsilon_2}$ (impedance of medium II).

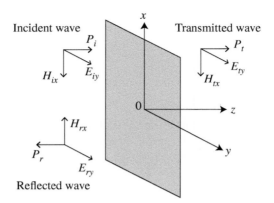

Figure 2.22 Electromagnetic field at the medium interface.

The configuration of these EM field components is shown in Figure 2.22.

When the electric field components of the incident, reflected, and transmitted waves exist in the positive y-axis direction, both the incident and the transmitted magnetic fields are oriented in the negative x-axis direction. Also, the reflected magnetic field component H_{rx} orients in the positive direction of the x axis.

The reason why the H_{rx} component takes the opposite direction against H_{ix} will be apparent from the Poynting vector viewpoint.

Now, to determine the unknown coefficients A, B, and C (amplitude) in Eqs. (2.94) and (2.98), the following boundary conditions should be imposed. That is, "the tangential components of the electric field and the magnetic field are equal to each other on both sides of the boundary surface."

$$E_{y1}|_{z=0} = E_{y2}|_{z=0} \tag{2.100}$$

$$H_{x1}|_{z=0} = H_{x2}|_{z=0}. \tag{2.101}$$

From Eqs. (2.94) and (2.98),

$$E_{y1} = E_{iy} + E_{ry} = E_{y2} = E_{ty} \quad (z = 0). \tag{2.102}$$

As for the magnetic fields, from Eqs. (2.95) and (2.99),

$$H_{x1} = H_{ix} + H_{rx} = H_{x2} = H_{tx} \quad (z = 0). \tag{2.103}$$

Substituting Eqs. (2.94), (2.95), (2.98), and (2.99) into Eqs. (2.102) and (2.103), the following relations can be derived.

$$A + B = C \tag{2.104}$$

$$A - B = (\eta_1/\eta_2)C. \tag{2.105}$$

By solving these simultaneous equations for the reflection coefficient $R = B/A$ and the transmission coefficient $T = C/A$, the following expressions are obtained:

$$\text{Reflection coefficient}: R = B/A = (\eta_2 - \eta_1)/(\eta_2 + \eta_1) \qquad (2.106)$$

$$\text{Transmission coefficient}: T = C/A = (2\eta_2)/(\eta_2 + \eta_1). \qquad (2.107)$$

Note that the same result is derived even in if the plane transverse magnetic (TM) wave propagates in the z-axis direction with the normal incidence on the boundary surface.

2.3.4.2 Oblique Incidence

Let us consider here the reflection and transmission characteristics when the plane TE and TM waves are obliquely incident on the boundary surface between the two mediums which spread infinitely. As before, the values of permittivity and permeability in medium I and II are represented by (ε_1, μ_1) and (ε_2, μ_2), respectively.

(a) TE Wave Case As can be seen from the previous section, in the case of a normal incident, the TE and TM waves result in the same expressions for the reflection coefficient and the transmission coefficient. On the contrary, in the case of oblique incidence, these relationships take different expressions depending on the TE and TM waves. First, the TE wave case is taken up. As is apparent from Eq. (2.94) in the previous section, the incident electric field propagating in the positive direction of the z axis is represented as follows.

$$E_{iy} = Ae^{-j\gamma_1 z}. \qquad (2.108)$$

In the oblique incidence, as shown in Figure 2.23, it is only natural that the direction of EM-wave propagation does not coincide with the coordinate axis z. The treatment in such a case can be resolved by rotating the coordinate axes.[1]

As shown in Figure 2.24, let us consider a new coordinate $(\bar{x}, \bar{y}, \bar{z})$ which is rotated by an angle θ from the z-axis. As a result, the relationship between the new coordinates and the old coordinates is

$$z = \bar{z} \cos \theta \quad x = -\bar{z} \sin \theta.$$

When using the formula $\cos^2\theta + \sin^2\theta = 1$ and expressing the given expression by $\bar{z}$,

$$\bar{z} = z \cos \theta - x \sin \theta. \qquad (2.109)$$

---

1 According to the principle of the right-handed coordinate system, the direction of rotation of the coordinate, generally, should be counterclockwise. However, in order to make it easier to understand the present theory, we will rotate here clockwise in accordance with the same direction of the actual incident wave.

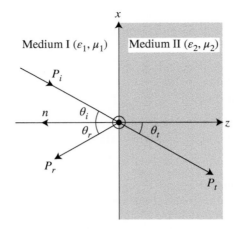

Figure 2.23 The case where a plane wave is obliquely incident on the boundary surface.

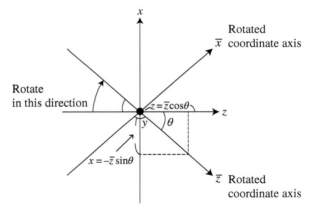

Figure 2.24 Rotation of coordinate axes.

Accordingly, the new incident electric field expression in the new coordinates $(\bar{x}, \bar{y}, \bar{z})$ can be represented as follows:

$$E_{iy}(\bar{z}) = Ae^{-j\gamma\bar{z}}. \tag{2.110}$$

With these preparations, let us derive the incident wave and the reflected wave in medium I, and also the transmitted wave in medium II.

Incident wave in medium I: Substituting Eq. (2.109) into Eq. (2.110), the incident electric field of medium I in the original coordinate system can be derived.

That is,

$$E_{iy}(x, z) = Ae^{-j\gamma_1(z\cos\theta_i - x\sin\theta_i)}. \qquad (2.111)$$

These are the expressions for the electric field component of the TE wave at oblique incidence. On the other hand, the magnetic field can be derived by substituting Eq. (2.111) into the relation of $H_x = -j/\omega\mu\partial E_y/\partial z$ in Eq. (2.93) in the previous section. That is,

$$
\begin{aligned}
H_{ix}(x, z) &= -j\frac{1}{\omega\mu_1}\frac{\partial E_{iy}(x, z)}{\partial z} \\
&= -\frac{A\cos\theta_i}{\eta_1}e^{-j\gamma_1(z\cos\theta_i - x\sin\theta_i)},
\end{aligned} \qquad (2.112)
$$

where $\eta_1 = \sqrt{\mu_1/\varepsilon_1}$.

Reflected wave in medium I: Next, for the reflected waves of the electric field and the magnetic field in medium I, they are easily obtained by reversing the sign of z in Eqs. (2.111) and (2.112).

$$E_{ry}(x, z) = Be^{j\gamma_1(z\cos\theta_r + x\sin\theta_r)} \qquad (2.113)$$

$$H_{rx}(x, z) = \frac{B\cos\theta_r}{\eta_1}e^{j\gamma_1(z\cos\theta_r + x\sin\theta_r)}. \qquad (2.114)$$

Transmitted wave to medium II: As for the transmitted wave to medium II, the electric field and the magnetic field can be found from the following equations obtained by replacing γ_1 with γ_2, with θ_i and θ_t, and A with C, in Eq. (2.111) and Eq. (2.112):

$$E_{ty}(x, z) = Ce^{-j\gamma_2(z\cos\theta_t - x\sin\theta_t)} \qquad (2.115)$$

$$
\begin{aligned}
H_{tx}(x, z) &= -j\frac{1}{\omega\mu_2}\frac{\partial E_{tx}}{\partial z} \\
&= -\frac{C\cos\theta_t}{\eta_2}e^{-j\gamma_x(z\cos\theta_t - x\sin\theta_t)}.
\end{aligned} \qquad (2.116)
$$

In the following, after obtaining simultaneous equations with unknown coefficients B and C, the boundary conditions are applied to determine B and C.

By considering the same way as in Section 2.3.4.1, the boundary conditions for the electric field and the magnetic field are imposed by the following equations:

$$E_{iy}(x, z)|_{z=0} + E_{ry}(x, z)|_{z=0} = E_{ty}(x, z)|_{z=0}$$
$$H_{ix}(x, z)|_{z=0} + H_{rx}(x, z)|_{z=0} = H_{tx}(x, z)|_{z=0}.$$

When applying these boundary conditions to Eqs. (2.111)–(2.116), the following equations are obtained:

$$Ae^{j\gamma_1 x \sin\theta_i} + Be^{j\gamma_1 x \sin\theta_r} = Ce^{j\gamma_2 x \sin\theta_t} \tag{2.117}$$

$$\frac{A\cos\theta_i}{\eta_1}e^{j\gamma_1 x \sin\theta_i} - \frac{B\cos\theta_r}{\eta_1}e^{j\gamma_1 x \sin\theta_r} = \frac{C\cos\theta_t}{\eta_2}e^{j\gamma_2 x \sin\theta_t}. \tag{2.118}$$

Here, when solving the simultaneous Eqs. (2.117) and (2.118) with the help of the assumed A as a known number and B and C as unknown numbers, we can obtain solutions for B and C as the functions of the variable x. That is, B and C cannot be determined as constant values.

In order to determine A and B as constant values, the exponential terms in both Eqs. (2.117) and (2.118) always have to be equal for any value on the x axis at the boundary $z = 0$.

This condition yields the following expression:

$$\gamma_1 x \sin\theta_i = \gamma_1 x \sin\theta_r = \gamma_2 x \sin\theta_t. \tag{2.119}$$

This condition is defined as the phase-matching condition. Based on this condition, the following important conditions can be derived:

$$\text{Law of reflection}: \theta_i = \theta_r \equiv \theta \tag{2.120}$$

$$\text{Snell's law}: \gamma_1 \sin\theta_i = \gamma_2 \sin\theta_t \tag{2.121}$$

or

$$\frac{\sin\theta_i}{\sin\theta_t} = \frac{\gamma_2}{\gamma_1} \equiv n.$$

Equation (2.120) means that the reflection and incident angles are equal, which is the law of reflection.

Also, Eq. (2.121) states that the ratio of the sine of the transmitted angle with respect to the incident angle is constant. This relation is called Snell's law.

By the way, the following expressions were obtained by solving expressions (2.117) and (2.118) under the conditions of expressions (2.120) and (2.121).

$$A + B = C \tag{2.122}$$

$$A - B = \frac{\eta_1 \cos\theta_t}{\eta_2 \cos\theta_i}C. \tag{2.123}$$

Solving these two equations, the reflection coefficient and the transmission coefficient can be derived as follows:

$$R_E = \frac{B}{A} = \frac{1 - \rho_E}{1 + \rho_E} \tag{2.124}$$

$$T_E = \frac{C}{A} = \frac{2}{1 + \rho_E}, \tag{2.125}$$

where

$$\rho_E = \frac{\eta_1 \cos\theta_t}{\eta_2 \cos\theta_i} = \frac{\mu_1 \sqrt{(\gamma_2/\gamma_1)^2 - \sin^2\theta_i}}{\mu_2 \cos\theta_i}.$$

Equation (2.124) is usually called the Fresnel's reflection coefficient.

(b) **TM Wave Case** Next, the task is to derive the reflection and transmission coefficients when the plane TM wave is obliquely incident on the boundary surface of two media extending infinitely as shown in Figure 2.25.

Upon applying the relations $H = jH_y$, $\partial/\partial x = 0$, and $\partial/\partial y = 0$ to Maxwell's equations, the present TM wave equation can be determined in the same way as for the TE wave.

The component of H_y is expressed by the following differential equation in the same way as in the previous case of the TE wave:

$$\frac{\partial^2 H_y}{\partial z^2} + \gamma^2 H_y = 0. \tag{2.126}$$

Hence,

$$H_y = A' e^{-j\gamma(z\cos\theta - x\sin\theta)}. \tag{2.127}$$

Also, the E_x component can be represented as

$$E_x = \frac{j}{\omega\varepsilon} \frac{\partial H_y}{\partial z}. \tag{2.128}$$

First, in the same way as the previous TE-wave case, let us derive the incident wave in medium I, and the reflected wave and the transmitted wave in medium II.

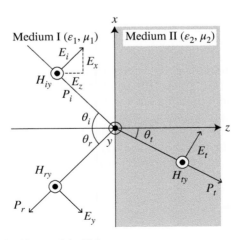

Figure 2.25 Oblique incidence of the TM wave.

Incident wave in medium I:

$$H_{iy} = A'e^{-j\gamma_1(z\cos\theta_i - x\sin\theta_i)} \tag{2.129}$$

$$E_{ix} = \eta_1 \cos\theta_i A'e^{-j\gamma_1(z\cos\theta_i - x\sin\theta_i)}. \tag{2.130}$$

Reflected wave in medium I: By reversing the sign of z in Eqs. (2.129) and (2.130)

$$H_{ry} = B'e^{j\gamma_1(z\cos\theta_r + x\sin\theta_r)} \tag{2.131}$$

$$E_{rx} = \eta_1 \cos\theta_r B'e^{j\gamma_1(z\cos\theta_r + x\sin\theta_r)}. \tag{2.132}$$

Transmitted wave in medium II: Replacing θ_i in Eq. (2.129) by θ_t, and γ_1 in Eq. (2.130) by γ_2, yields

$$H_{ty} = C'e^{-j\gamma_2(z\cos\theta_t - x\sin\theta_t)} \tag{2.133}$$

$$E_{tx} = \eta_2 \cos\theta_1 C'e^{-j\gamma_2(z\cos\theta_t - x\sin\theta_t)}. \tag{2.134}$$

In the present case, only each of the tangential components of the electric field E_x and the magnetic field H_y are involved in the boundary conditions. Namely,

$$H_{iy}|_{z=0} = H_{ty}|_{z=0}$$
$$E_{ix}|_{z=0} = E_{tx}|_{z=0}. \tag{2.135}$$

In the same way as in the previous TE-wave case, the present boundary conditions are represented by the following equations:

$$H_{iy}(x,z)|_{z=0} + H_{ry}(x,z)|_{z=0} = H_{ty}(x,z)|_{z=0}$$
$$E_{ix}(x,z)|_{z=0} + E_{yx}(x,z)|_{z=0} = E_{tx}(x,z)|_{z=0}.$$

Simultaneous equations can be expressed by assuming a known number A' and unknowns B' and C', using the relations of Eqs. (2.129)–(2.134) and the given boundary condition in the same way as in the TE-wave case.

Hence, after some calculations, the A', B', and C' coefficient values are obtained, which are necessary for calculating the reflection coefficient and the transmission coefficient, in the similar way of TE-wave.

Consequently, the reflection coefficient R_M and the transmission coefficient T_M are given.

Reflection coefficient:

$$R_M = \frac{B'}{A'} = \frac{1 - \rho_M}{1 + \rho_M} \tag{2.136}$$

Transmission coefficient:

$$T_M = \frac{C'}{A'} = \frac{2}{1 + \rho_M},$$

(2.137)

where $\rho_M = \dfrac{\eta_2 \cos \theta_t}{\eta_1 \cos \theta_i} = \dfrac{\varepsilon_1 \sqrt{(\gamma_2/\gamma_1)^2 - \sin^2 \theta_i}}{\varepsilon_2 \cos \theta_i}.$

From these investigations of reflection and transmission coefficients, it can be confirmed that there exists only a difference in the permittivity and permeability value between the TM wave and the TE wave.

2.3.5 Theory of Multiple Reflections

2.3.5.1 Reflection and Transmission Coefficients

It is important to deepen the theory of multiple reflections, which is the basis for studying the EM-wave absorber problems.

As shown in Figure 2.26, let us consider the case when a plane EM wave is perpendicularly incident from medium I into medium II, which consists of the dielectric constant and the magnetic permeability.

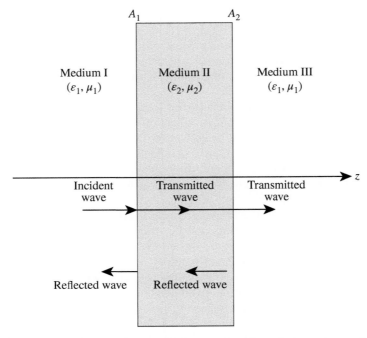

Figure 2.26 Reflection and transmission of the normal incidence in the presence of media with different electric constants.

Mediums I and III have the same permittivity (ε_1) and permeability (μ_1), respectively. For simplicity, we here take up the case where each medium is lossless.

Now, when a plane wave vertically incidents on this medium II, the reflection coefficient and the transmission coefficient at the boundary surface A_1 are expressed by the following equations, when the impedance of each medium is denoted by Z_1 and Z_2.

At the boundary surface A_1,

$$\text{The reflection coefficient}: R = \frac{Z_2 - Z_1}{Z_2 + Z_1} \tag{2.138}$$

$$\text{The transmitted coefficient}: T = \frac{2Z_2}{Z_1 + Z_2} = 1 + R. \tag{2.139}$$

In addition, on the boundary surface A_2, the reflection coefficient and the transmission coefficient at the interface A_2 are given by the following equations taking into account the positional relationship of the medium, which is opposite of that in the case of A_1.

At the boundary surface A_2,

$$\text{The reflection coefficient}: R' = \frac{Z_1 - Z_2}{Z_1 + Z_2} = -R \tag{2.140}$$

$$\text{The transmitted coefficient}: T' = 1 + R' = 1 - R. \tag{2.141}$$

Now it becomes possible to analyze the multiple reflection problems using the relations of the given equations. When the incident EM-wave amplitude on the boundary line A_1 is assumed to be 1 for simplicity, this wave can be denoted by e^{-jk_1z}.

In Figure 2.27, for the sake of clarity, notice that this figure is depicted as if the oblique incident and the oblique reflection waves exist. Well, the incident wave ① is first reflected on the boundary surface A_1, and this reflected wave is expressed as Re^{+jk_1z} in terms of the reflection coefficient R. This means the primary reflected waves ②, and the reflection coefficient is given by Eq. (2.138). Moreover, the incident wave ① propagates in the medium II with the thickness d.

At that time, the transmitted wave ③ causes the phase lag which is expressed as e^{-jk_2d} at A_2. Therefore, multiplying the transmission coefficient T in Eq. (2.139), the transmitted wave ③ can be denoted by $E_③^T = Te^{-jk_2d}$ at A_2.

In addition, this $E_③^T$ becomes an incident wave for the boundary surface A_2 and is reflected on the boundary line A_2. This wave again propagates to the A_1 boundary direction as a reflected wave against the direction of transmitted wave $E_③^T$. For this reflected wave ⑤, multiplying the reflection coefficient at the interface A_2 is expressed in Eq. (2.140), and then, taking into account the phase delay against $E_③^T$, this can be expressed as $E_⑤^R = E_③^T R' e^{-jk_2d}$ on the boundary line A_1. Next, by multiplying the transmission coefficient

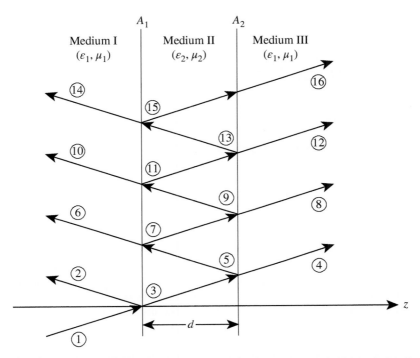

Figure 2.27 Illustration of multiple reflected waves and transmitted waves. (For clarity of the present explanation, this figure is depicted as if it were an oblique incidence case. In practice, vertical incidence occurs, vertical reflections, and transmissions are repeated.)

in Eq. (2.141) by $E_⑤^R$, the transmitted wave ⑥ from medium II to I can be expressed as $E_⑥^T = T'E_⑤^R = TT'R'e^{-j2k_2d}$. This ER⑥ in medium I looks like a reflected wave (referred to as first echo) from the boundary A_1. Hence, E⑥ can be regarded as a reflected wave.

Next, let us consider the wave ④ at the boundary A_2 which has passed through to medium III. This equation can be determined by multiplying the T' in Eq. (2.141) by $E_③^T$ in medium II.

That is, $E_④^T = T'E_③^T = T'Te^{-jk_2d}$.

If these procedures were repeatedly applied on each boundary surface in A_1 and A_2, by taking all the sums about each reflection coefficient and each transmission coefficient, we can derive the total reflection coefficient and the total transmission coefficient. That is, the reflection coefficient is in front of boundary A_1

$$S = Re^{jk_1z} + TR'T'e^{-j2k_2d} + TR'^3T'e^{-j4k_2d} + \cdots$$
$$= R\left(1 - \frac{TR'T'e^{-j2k_2d}}{1 - R^2e^{-j2k_2d}}\right) = \frac{R(1 - e^{-j2k_2d})}{1 - R^2e^{-j2k_2d}}. \tag{2.142}$$

Similarly, the transmission coefficient is at the boundary A_2

$$T = TT'e^{-jk_2d} + TR'^2T'e^{-j3k_2d} + TR'^4T'e^{-j5k_2d} + \cdots$$
$$= \frac{TT'e^{-jk_2d}}{1 - R^2e^{-j2k_2d}} = \frac{(1 - R^2)e^{-j2k_2d+jk_1d}}{1 - R^2e^{-j2k_2d}}. \tag{2.143}$$

2.A Appendix

2.A.1 Appendix to Section 2.3.2 (1)

Explanation of magnetic current density:
The following equation represents Ampere's circuital law.

$$\oint_C H \cdot dl = I. \tag{2.A.1}$$

This means that a loop-shaped magnetic field H is induced by the current I as shown in Figure 2.A.1a. dl is a unit vector having a length derived by finely dividing the closed loop and a tangential direction at that point. It should be noted that the current I in this case consists of the conduction current density J flowing through the conductor and the displacement current density $\partial D/\partial t$. By analogy with this relationship in Eq. (2.A.1), if the electric field E is integrated along the closed loop, as shown in Figure 2.A.1b, this coincides with Faraday's law being equal to the time change of the interlinking magnetic flux, and, consequently, the following equation is held:

$$\oint_C E \cdot dl = -\frac{\partial \varphi}{\partial t} = -M. \tag{2.A.2}$$

$\partial \varphi/\partial t = M$ in this Eq. (2.A.2) is found to be corresponding to the current I in Eq. (2.A.1). Therefore, $\partial \varphi/\partial t = M$ can be regarded as a magnetic current with respect to the current shown in Eq. (2.A.1). This means that the current I is given by the closed circuital integration of the magnetic field and also the magnetic current M is represented by the closed circuital integration of the electric field. The expression incorporating the concept of this magnetic

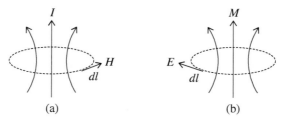

(a)   (b)

Figure 2.A.1 Explanation of magnetic current. (a) The law of Ampere's circuital law and (b) the magnetic current density M.

current is the right side of the expression (2.72). $\partial \boldsymbol{B}/\partial t$ is the displacement magnetic current density with time varying, and $\boldsymbol{M}$ is the magnetic current density. These have the dimension of voltage.

References

1 Dawson, J.F. (1993). Representing ferrite absorber tiles as frequency dependent boundaries in TLM. *Electron. Lett.* 29 (9): 289–290.

2 Chambers, B. (1997). Dynamically adaptive radar absorbing material with improved self-monitoring characteristics. *Electron. Lett.* 33 (6): 529–530.

3 Suetake, K. and Hayashi, S. (1970). *Microwave Circuit*. Ohmsha LTD.

4 Amano, M. and Kotsuka, Y. (2015). Detailed investigations on flat single layer selective magnetic absorber based on the Equivalent Transformation method of material constants. *IEEE Trans. Electromagn. Compat.* 57 (6): 1398–1407.

5 Suetake, K. (1952). Gum transformation - mechanical model of conformal mapping and its application. *Trans. IECE*, Japan 35 (12): 573–575.

3

Methods of Absorber Analysis

The pyramidal absorbers described in Chapter 1 in the classification of electromagnetic (EM)-wave absorber appearance have been extensively used in anechoic chambers and have already reached sufficient technical achievement levels. In addition, various design methods have been proposed so far regarding these pyramidal absorbers, and their design methods have already been broadly established from aspects of both analysis and actual measurement, as discussed in Chapter 1 [1–5]. On the other hand, various application fields have been expected for flat-type absorbers along with the rapid progress of future communication environments. In view of these situations, in this book we mainly focus on single-layer plate-type absorbers and a multilayer-type absorber composed of flat plate absorbers.

Sections 3.1 and 3.2 describe how to analyze the absorbing characteristics when the plane wave is incident normally or obliquely to single-layer-type absorbers. In Section 3.3, the absorbing characteristics when a plane wave is incident on a multilayer flat-type absorber normally or obliquely are treated.

Furthermore, in Section 3.4, as an example of the way of thinking of multiple reflections, one of the theoretical analysis methods is concretely introduced, taking an example of multi-reflection problems in the case of an EM-wave absorber placed in a measurement room.

3.1 Normal Incidence to Single-layer Flat Absorber

We shall first try to investigate the reflection coefficient of a single-layer flat absorber backed with a conductive plate on the basis of the fundamental theory in Chapter 2. As shown in Figure 3.1, medium I consists of air (free space treatment), and medium II is a single-layer absorber, having the boundary surface at $z = 0$. The thickness of the EM-wave absorber is d, and the conductive plate is attached at the position of $z = d$ on its back.

The plane normal incident EM wave propagates from medium I to medium II. Since the transmitted wave to medium II is reflected by the conductive plate,

Electromagnetic Wave Absorbers: Detailed Theories and Applications, First Edition. Youji Kotsuka.
© 2019 John Wiley & Sons, Inc. Published 2019 by John Wiley & Sons, Inc.

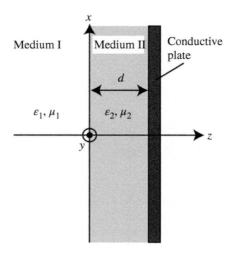

Figure 3.1 Single-layer EM-wave absorber.

the appearance of a standing wave can be considered in medium II. Hence, for the electric field of medium II, we can formally use the form of expression (2.87) in Section 2.3.3, but it should be represented so as to make the propagation constant and material constants corresponding to medium II. For convenience, let us write down here only the x and y components of the electromagnetic fields in each medium.

Medium I:

$$E_{y1}(z) = Ae^{-j\gamma_1 z} + Be^{j\gamma_1 z} \tag{3.1}$$

$$H_{x1}(z) = -\frac{1}{Z_{c1}}(Ae^{-j\gamma_1 z} - Be^{j\gamma_1 z}) \tag{3.2}$$

Medium II:

$$E_{y2}(z) = Ce^{-j\gamma_2 z} + De^{j\gamma_2 z} \tag{3.3}$$

$$H_{x2}(z) = -\frac{1}{Z_{c2}}(Ce^{-j\gamma_2 z} - De^{j\gamma_2 z}), \tag{3.4}$$

where Z_{c1}, Z_{c2} represent the characteristic impedance in each medium I and II.

$$\gamma_1 = \omega\sqrt{\varepsilon_1\mu_1} \tag{3.5}$$

$$\gamma_2 = \omega\sqrt{\varepsilon_2\mu_2}. \tag{3.6}$$

In the present case, the tangential components of the electric and magnetic fields should be taken into account as boundary conditions in order to be equal at the boundary $z = 0$ between medium I and II. And then, the tangential component of the electric field on the conductor surface of the $z = d$ must be zero.

That is, from the relations in equation from (3.1) to (3.4)

$$z = 0 \quad E_{y1}|_{z=0} = E_{y2}|_{z=0} \tag{3.7}$$

$$H_{x1}|_{z=0} = H_{x2}|_{z=0} \tag{3.8}$$

$$z = d \quad E_{y2}|_{z=d} = 0. \tag{3.9}$$

First, from Eq. (3.7)

$$A + B = C + D. \tag{3.10}$$

In addition, from Eq. (3.8),

$$-\frac{1}{Z_{c1}}(A - B) = -\frac{1}{Z_{c2}}(C - D) \quad \therefore A - B = \frac{Z_{c1}}{Z_{c2}} C \left(1 - \frac{D}{C}\right). \tag{3.11}$$

From the condition of Eq. (3.9),

$$\frac{D}{C} = -\frac{e^{-j\gamma_2 d}}{e^{j\gamma_2 d}}. \tag{3.12}$$

Here, substituting Eq. (3.12) into Eqs. (3.10) and (3.11) and taking both the sum and the difference of Eqs. (3.10) and (3.11), the following expressions are derived:

$$2A = C\left\{1 - \frac{e^{-j\gamma_2 d}}{e^{j\gamma_2 d}}\right\} + C\frac{Z_{c1}}{Z_{c2}}\left\{1 + \frac{e^{-j\gamma_2 d}}{e^{j\gamma_2 d}}\right\} \tag{3.13}$$

$$2B = C\left\{1 - \frac{e^{-j\gamma_2 d}}{e^{j\gamma_2 d}}\right\} - C\frac{Z_{c1}}{Z_{c2}}\left\{1 + \frac{e^{-j\gamma_2 d}}{e^{j\gamma_2 d}}\right\}. \tag{3.14}$$

Thus, the reflection coefficient can be expressed by the following equation on the interface of medium I and II.

$$S = \frac{B}{A} = \frac{\dfrac{Z_{c2}}{Z_{c1}}\{e^{j\gamma_2 d} - e^{-j\gamma_2 d}\} - \{e^{j\gamma_2 d} + e^{-j\gamma_2 d}\}}{\dfrac{Z_{c2}}{Z_{c1}}\{e^{j\gamma_2 d} - e^{-j\gamma_2 d}\} + \{e^{j\gamma_2 d} + e^{-j\gamma_2 d}\}}$$

$$= \frac{\dfrac{Z_{c2}}{Z_{c1}}\dfrac{e^{j\gamma_2 d} - e^{-j\gamma_2 d}}{e^{j\gamma_2 d} + e^{-j\gamma_2 d}} - 1}{\dfrac{Z_{c2}}{Z_{c1}}\dfrac{e^{j\gamma_2 d} - e^{-j\gamma_2 d}}{e^{j\gamma_2 d} + e^{-j\gamma_2 d}} + 1}$$

$$= \frac{\dfrac{Z_{c2}}{Z_{c1}}\tanh j\gamma_2 d - 1}{\dfrac{Z_{c2}}{Z_{c1}}\tanh j\gamma_2 d + 1}. \tag{3.15}$$

Now, since medium I is assumed to be a free space, the wavelength is λ_0, and the characteristic impedance Z_{c1} in medium I is,

$$Z_{c1} = \sqrt{\frac{\mu_0}{\varepsilon_0}} = Z_0.$$

On the other hand, in medium II, the impedance is generally given by

$$Z_{c2} = \sqrt{\frac{\mu_2}{\varepsilon_2}} = Z_0 \sqrt{\frac{\mu_{r2}}{r_{\gamma 2}}}.$$

Consequently, the reflection coefficient S is

$$S = \frac{\sqrt{\frac{\mu_{r2}}{\varepsilon_{r2}}} \tanh\left(j\frac{2\pi}{\lambda_0}\sqrt{\varepsilon_{r2}\mu_{r2}}d\right) - 1}{\sqrt{\frac{\mu_{r2}}{\varepsilon_{r2}}} \tanh\left(j\frac{2\pi}{\lambda}\sqrt{\varepsilon_{r2}\mu_{r2}}d\right) + 1}. \tag{3.16}$$

3.2 Oblique Incidence to Single-layer Flat Absorber

In this section, we investigate how to derive the reflection coefficient in the case when the TE plane wave is obliquely incident onto a single-layer wave absorber of thickness d.

Now the coordinate system is defined, as shown in Figure 3.2. In a typical wave absorber, medium I is treated as a free space, and medium II consists of any uniform absorbing material which has material constants.

The conductive plate is attached to the back surface of a uniform absorbing material. When the plane TE wave is incident on the EM-wave absorber with an incident angle θ_i to the z-axis, as shown in Figure 3.2, the traveling waves and the reflected wave coexist in medium II.

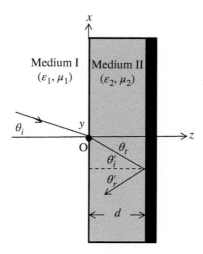

Figure 3.2 Configuration of the EM wave obliquely incident on the single-layer EM-wave absorber.

The electric field and magnetic field of each medium in I and II are expressed by the following equations, in the same way as in Section 2.3.4.2.

Medium I:

$$E_{y1} = -\frac{A\cos\theta_i}{Z_{c1}}e^{-j\gamma_1(z\cos\theta_i - x\sin\theta_i)} + \frac{B\cos\theta_r}{Z_{c1}}e^{j\gamma_1(z\cos\theta_r + x\sin\theta_r)} \tag{3.17}$$

$$H_{x1} = -\frac{A\cos\theta_i}{Z_{c1}}e^{-j\gamma_1(z\cos\theta_i - x\sin\theta_i)} + \frac{B\cos\theta_r}{Z_{c1}}e^{j\gamma_1(z\cos\theta_r + x\sin\theta_r)}. \tag{3.18}$$

Medium II:

$$E_{y2} = Ce^{-j\gamma_2(z\cos\theta_i' - x\sin\theta_i')} + De^{j\gamma_2(z\cos\theta_r' + x\sin\theta_r')} \tag{3.19}$$

$$H_{x2} = -\frac{C\cos\theta_i'}{Z_{c2}}e^{-j\gamma_2(z\cos\theta_i' - x\sin\theta_i')} + \frac{D\cos\theta_r'}{Z_{c2}}e^{j\gamma_2(z\cos\theta_r' + x\sin\theta_r')}, \tag{3.20}$$

θ_i', θ_r' denote the incident and reflected angle with respect to the conductor plate surface in medium II, but these angles are equal to the angle θ_t, based on the reflection law described in Section 2.3.4.2. Let us here think about imposing the following boundary conditions in Eqs. (3.17)–(3.20). That is, at $Z = 0$,

$$E_{y1}|_{z0} = E_{y2}|_{z0} \tag{3.21}$$

$$H_{x1}|_{z0} = H_{x2}|_{z0}. \tag{3.22}$$

And, at $Z = d$,

$$E_{y2}|_{z0} = 0. \tag{3.23}$$

Hence, from the boundary condition in (3.21)

$$Ae^{j\gamma_1 x\sin\theta_i} + Be^{j\gamma_1 x\sin\theta_r} = Ce^{j\gamma_2 x\sin\theta_i'} + De^{j\gamma_2 x\sin\theta_r'}. \tag{3.24}$$

Also, from the condition in (3.22),

$$Ae^{j\gamma_1 x\sin\theta_i} - Be^{j\gamma_1 x\sin\theta_r} = \frac{Z_{c1}\cos\theta_r'}{Z_{c2}\cos\theta_i'}C\left\{e^{j\gamma_2 x\sin\theta_i'} - \frac{D}{C}e^{j\gamma_2 x\sin\theta_r'}\right\}. \tag{3.25}$$

Further, from the condition in (3.23)

$$\frac{D}{C} = -\frac{e^{-j\gamma_2 d\cos\theta_i'}}{e^{j\gamma_2 d\cos\theta_r'}}. \tag{3.26}$$

Here, using the relationship $\theta_i' = \theta_r' = \theta_t$, and taking both the sum and difference between Eqs. (3.24) and (3.25), respectively, the following expressions can be derived:

$$2Ae^{j\gamma_1 x\sin\theta_i} = Ce^{j\gamma_2 x\sin\theta}\left(1 + \frac{D}{C}\right) + C\frac{Z_{c1}\cos\theta_t}{Z_{c2}\cos\theta_i}e^{j\gamma_2 x\sin\theta_t}\left(1 - \frac{D}{C}\right) \tag{3.27}$$

$$2Be^{j\gamma_1 x\sin\theta_r} = Ce^{j\gamma_2 x\sin\theta_t}\left(1 + \frac{D}{C}\right) - C\frac{Z_{c1}\cos\theta_t}{Z_{c2}\cos\theta_i}e^{j\gamma_2 x\sin\theta_t}\left(1 - \frac{D}{C}\right). \tag{3.28}$$

Noting the relationship $\theta_i = \theta_r$, after obtaining the ratio of B/A from Eqs. (3.27) and (3.28), and by substituting Eq. (3.26) into B/A, the reflection coefficient can be finally derived.

$$S = \frac{B}{A} = \frac{\left(1 + \frac{D}{C}\right) - \frac{Z_{c1}\cos\theta_t}{Z_{c2}\cos\theta_i}\left(1 - \frac{D}{C}\right)}{\left(1 + \frac{D}{C}\right) + \frac{Z_{c1}\cos\theta_t}{Z_{c2}\cos\theta_i}\left(1 - \frac{D}{C}\right)} = \frac{\frac{Z_{c2}\cos\theta_i}{Z_{c1}\cos\theta_t}\left(\frac{e^{j\gamma_2 d\cos\theta_t} - e^{-j\gamma_2 d\cos\theta_t}}{e^{j\gamma_2 d\cos\theta_t} + e^{-j\gamma_2 d\cos\theta_t}}\right) - 1}{\frac{Z_{c2}\cos\theta_i}{Z_{c1}\cos\theta_t}\left(\frac{e^{j\gamma_2 d\cos\theta_t} - e^{-j\gamma_2 d\cos\theta_t}}{e^{j\gamma_2 d\cos\theta_t} + e^{-j\gamma_2 d\cos\theta_t}}\right) + 1}.$$

(3.29)

Further, by applying the relations of the hyperbolic function described here, Eq. (3.29) is deformed as follows:

$$\sinh jx = \frac{e^{jx} - e^{-jx}}{2} = j\sin x$$

$$\cosh jx = \frac{e^{jx} + e^{-jx}}{2} = \cos x$$

$$\tanh jx = j\sin x / \cos x = j\tan x.$$

Then,

$$S = \frac{\frac{Z_{c2}\cos\theta_i}{Z_{c1}\cos\theta_t}(\tanh j\gamma_2 d\cos\theta_t) - 1}{\frac{Z_{c2}\cos\theta_i}{Z_{c1}\cos\theta_t}(\tanh j\gamma_2 d\cos\theta_t) + 1}.$$

(3.30)

From Snell's law,

$$\sin\theta_t = \frac{\gamma_1}{\gamma_2}\sin\theta_i.$$

Hence,

$$\therefore \cos\theta_t = \sqrt{1 - \left(\gamma_1/\gamma_2\right)^2\sin^2\theta_i}$$

$$= \sqrt{1 - \left(\frac{\varepsilon_0\mu_0}{\varepsilon_0\varepsilon_{r2}\mu_0\mu_{r2}}\right)^2\sin^2\theta_i}$$

$$= \sqrt{1 - \frac{1}{\varepsilon_{r2}\mu_{r2}}\sin^2\theta_i}.$$

(3.31)

In addition, the characteristic impedances of medium I and medium II are denoted as follows, respectively:

$$Z_{c1} = \sqrt{\frac{\mu_0}{\varepsilon_0}} = Z_0, \quad Z_{c2} = \sqrt{\frac{\mu_0\mu_{r2}}{\varepsilon_0\varepsilon_{r2}}} = Z_0\sqrt{\frac{\mu_{r2}}{\varepsilon_{r2}}}.$$

(3.32)

Therefore, applying Eqs. (3.31) and (3.32) to Eq. (3.30), yields

$$\frac{Z_{c2}\cos\theta_i}{Z_{c1}\cos\theta_t} = \sqrt{\frac{\mu_{r2}}{\varepsilon_{r2}}}\frac{\cos\theta_i}{\sqrt{1 - 1/\varepsilon_{r2}\mu_{r2}\sin^2\theta_i}} = \frac{\mu_{r2}\cos\theta_i}{\sqrt{\varepsilon_{r2}\mu_{r2} - \sin^2\theta_i}}.$$

(3.33)

Hence, Eq. (3.30) can be rewritten as follows:

$$S = \frac{\frac{\mu_{r2}\cos\theta_i}{\sqrt{\varepsilon_{r2}\mu_{r2}-\sin^2\theta_i}}\{\tanh(j\omega\sqrt{\varepsilon_0\varepsilon_{r2}\mu_0\mu_{r2}}\sqrt{1-1/\varepsilon_{r2}\mu_{r2}\sin^2\theta_i}d)\}-1}{\frac{\mu_{r2}\cos\theta_i}{\sqrt{\varepsilon_{r2}\mu_{r2}-\sin^2\theta_i}}\{\tanh(j\omega\sqrt{\varepsilon_0\varepsilon_{r2}\mu_0\mu_{r2}}\sqrt{1-1/\varepsilon_{r2}\mu_{r2}\sin^2\theta_i}d)\}+1}.$$

Further,

$$S = \frac{\frac{\mu_{r2}\cos\theta_i}{\sqrt{\varepsilon_{r2}\mu_{r2}-\sin^2\theta_i}}\tanh\left(j\frac{2\pi d}{\lambda_0}\sqrt{\varepsilon_{r2}\mu_{r2}-\sin^2\theta_i}\right)-1}{\frac{\mu_{r2}\cos\theta_i}{\sqrt{\varepsilon_{r2}\mu_{r2}-\sin^2\theta_i}}\tanh\left(j\frac{2\pi d}{\lambda_0}\sqrt{\varepsilon_{r2}\mu_{r2}-\sin^2\theta_i}\right)+1}.$$

The final expression of the reflection coefficient in the case of the oblique incidence to a single-layer flat absorber is

$$S = \frac{Z_{\text{TE}}-\frac{1}{\cos\theta_i}}{Z_{\text{TE}}+\frac{1}{\cos\theta_i}}, \tag{3.34}$$

where

$$Z_{\text{TE}} = \frac{\mu_{r2}}{\sqrt{\varepsilon_{r2}\mu_{r2}-\sin\theta_i}}\tanh\left(j\frac{2\pi d}{\lambda_0}\sqrt{\varepsilon_{r2}\mu_{r2}-\sin^2\theta_i}\right), \tag{3.35}$$

where ε_{r2}, μ_{r2} represent the relative permittivity and relative permeability of medium II, respectively.

By the way, when a plane wave is incident perpendicularly to a single-layer absorber, of course, θ_i can be set to 0 in Eqs. (3.34) and (3.35).

As a matter of course, the reflection coefficient coincides with Eq. (3.16) in the previous section.

As a result,

$$S = \frac{\sqrt{\frac{\mu_{r2}}{\varepsilon_{r2}}}\tanh\left(j\frac{2\pi d}{\lambda_0}\sqrt{\varepsilon_{r2}\mu_{r2}}\right)-1}{\sqrt{\frac{\mu_{r2}}{\varepsilon_{r2}}}\tanh\left(j\frac{2\pi d}{\lambda_0}\sqrt{\varepsilon_{r2}\mu_{r2}}\right)+1}. \tag{3.16}$$

In the case of the TM wave, the reflection coefficient can be derived similarly.

3.3 Characteristics of the Multilayered Absorber

3.3.1 Normal Incidence Case

In this section, we deal with the method of obtaining the reflection coefficient in a "multilayer-type absorber" [6]. The multilayer EM-wave absorber

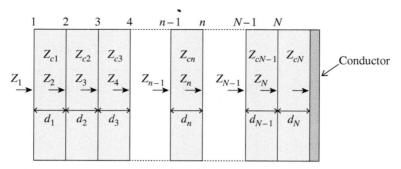

Figure 3.3 Multilayer EM-wave absorber. This simple and excellent analysis method was proposed by Professor Yasutaka Shimizu of the Tokyo Institute of Technology [6].

configuration, shown in Figure 3.3, can be constituted by each medium composed of an absorption layer with different material constants and thicknesses.

Generally, this kind of absorber is aimed at the realization of broadband frequency characteristics.

By the way, here we treat the problem on the basis of the transmission line theory, which is equivalent to the present EM-wave absorber configuration.

We shall consider the case when a plane wave is vertically incident on the multilayered absorber surface from the left side, as shown in Figure 3.3.

Let us now express the characteristic impedance and the propagation constant in the nth layer in terms of Z_{cn} and γ_n, respectively.

$$Z_{cn} = Z_0 \sqrt{\frac{\mu_{rn}}{\varepsilon_{rn}}} = Z_0 \sqrt{\frac{\mu'_{rn} - j\mu''_{rn}}{\varepsilon'_{rn} - j\varepsilon''_{rn}}} \tag{3.36}$$

$$\gamma_n = j\frac{2\pi}{\lambda_0}\sqrt{\varepsilon_{rn}\mu_{rn}} = j\frac{2\pi}{\lambda_0}\sqrt{(\varepsilon'_{rn} - j\varepsilon''_{rn})(\mu'_{rn} - j\mu''_{rn})}. \tag{3.37}$$

Here, ε_{rn} and μ_{rn} represent a complex relative permittivity and a complex relative permeability in the nth layer, respectively. Now, if the impedance looking at the front surface of the multilayer absorber is known, that is, if the input impedance against the boundary surface 1 can be determined, the reflection coefficient can be represented in terms of the following equation derived in Chapter 2, Eq. (2.32).

$$z_l = \frac{Z_l}{Z_c} = \frac{z_R + \tanh \gamma l}{1 + z_R \tanh \gamma l}, \tag{2.32}$$

where Z_0 is the free space impedance given by $Z_0 = 120\pi[\Omega]$.

By the way, after determining the input impedance expression Z_{n-1} with respect to the nth layer, the input impedance Z_1 can be calculated using the recurrence equation.

Referring to Eq. (2.32), this recurrence equation is expressed as follows:

$$Z_{n-1} = Z_{cn-1} \frac{Z_n + Z_{cn-1} \tanh \gamma_{n-1} d_{n-1}}{Z_{cn-1} + Z_n \tanh \gamma_{n-1} d_{n-1}}, \tag{3.38}$$

where $n = (N, N - 1, \dots, 3, 2)$. (Note here the difference between the upper numbers and the interlayer numbers depicted in Figure 3.3. The impedance number Z_n means an interlayer number in the present case.)

The input impedance Z_N for the last single-layer absorber in the Nth layer is given by the following equation:

$$Z_N = Z_{CN} \tanh \gamma_N d_N = Z_{CN} \tanh \left(j\frac{2\pi}{\lambda} \sqrt{\varepsilon_{rN} \mu_{rN}} d_N \right). \tag{3.39}$$

To obtain the input impedance Z_1 in Eq. (3.38), we have to do the following procedure.

This recurrence equation in (3.38) should be calculated from the terminating conductor plate side and the calculation is repeated until the input impedance value becomes Z_1 sequentially from N.

Consequently, since Z_1 corresponds to the load impedance for looking into the absorber surface of the first layer, the reflection coefficient $S(= (Z_1 - Z_0)/(Z_1 + Z_0))$ can be calculated by this Z_1 and the characteristic impedance Z_0 of the free space.

3.3.2 Case of Oblique Incidence

Even when a plane EM wave is obliquely incident on a multilayer-type absorber, the reflection coefficient can be calculated using the recurrence Eq. (3.38) in the previous section. It should be noted, however, that the formulae for calculating the characteristic impedance Z_{cn}, and reflection coefficient S in the nth layer become different, depending on the TM or TE wave, as follows.

3.3.2.1 Case of the TE Wave

$$Z_{cn} = \frac{Z_0 \mu_{rn}}{\sqrt{\varepsilon_{rn} \mu_{rn} - \sin^2 \theta}} \tag{3.40}$$

$$\gamma_n = j\frac{2\pi}{\lambda_0} \sqrt{\varepsilon_{rn} \mu_{rn} - \sin^2 \theta} \tag{3.41}$$

$$S_{TE} = \frac{Z_1 - Z_0/\cos\theta}{Z_1 + Z_0/\cos\theta}. \tag{3.42}$$

3.3.2.2 Case of the TM Wave

$$Z_{cn} = \frac{Z_0 \sqrt{\varepsilon_{rn} \mu_{rn} - \sin^2 \theta}}{\varepsilon_{rn}} \tag{3.43}$$

$$\gamma_n = j\frac{2\pi}{\lambda_0}\sqrt{\varepsilon_{rn}\mu_{rn} - \sin^2\theta} \tag{3.44}$$

$$S_{\text{TM}} = \frac{Z_1 - Z_0\cos\theta}{Z_1 + Z_0\cos\theta}. \tag{3.45}$$

3.4 Case of Multiple Reflected and Scattered Waves

The theory described until now was treated as the case where the size of the absorber was of an infinitely large size.

In contrast, the EM-wave absorbers used in an anechoic chamber and a simple measurement site have finite length. In this section, we discuss the scattering problems in the case of setting a pyramid-type for measuring this kind of wave absorber with finite dimensions in an indoor place [6, 7]. Currently, the network analyzer has such functions as the time-domain method, and its performance is improved. But the objective in this section is to comprehensively understand the mutual relationship between the incident wave, the reflected wave, and the scattered wave, which is the main concept of the EM-wave absorber research and deepens our knowledge of the means of visualization of reflection and scattering phenomena. In particular, how to treat the standing-wave ratio in the case of multiple reflections that arise is discussed using the example of the appearance of a beat wave.

The topics covered here also have significance from the viewpoint of the method for measuring EM-wave absorbers, which are related to the topics in Chapter 7. Thus, here we treat the problem in an environment that coexists with incident waves propagating obliquely to the direction of the pyramidal absorber and reflected waves from its absorber wall surfaces, and the like. When considering the multiple scattering wave problems of such an environment, the following wave appears on the front of the wave absorber.

The main ones are the occurrence of beat, the occurrence of undulation, and the case where beat and undulation coexist.

In this section, we describe the treatment of each equation of the incident electromagnetic field, the scattered electromagnetic field, and the idea of a standing-wave ratio, taking as an example the case of a beat.

Note here that the beat is a phenomenon where two waves with slightly different frequencies interfere with each other, resulting in a composite wave, the amplitude of which varies periodically.

For analytical treatment in cases of undulation, beat, and undulation mixed with beat, see Reference [7].

The setting conditions of the present beat theme are as follows:

(a) Electromagnetic waves from the wave transmission antenna are obliquely incident on a pyramidal-type absorber.

(b) Assume that the measuring instruments in this place are covered with EM-wave absorbers.

(c) As for the scattering waves, when the incident electric field from the transmitting antenna is E_1, the scattering electric fields from the EM-wave absorber are denoted by $(E_{s1}'', E_{s2}'', E_{s3}'', \ldots)$, but are assumed to be very weak compared to E_1.

(d) Assume that these incident waves, scattered waves, and reflected waves exist only in the y–z plane.

(e) These electric fields, which are treated here, take into account the directivity factor of the antenna.

(f) Here we examine only the case of a TE wave.

(g) In order to make the explanation easy to understand without covering all the various reflected waves and scattering, they are investigated with a focus on reflected waves and scattered waves of maximum intensities among them.

(h) All reflected waves and scattered waves, including the incident wave from the transmitter, are obtained by moving the standing-wave detection antenna placed from the left side on the y axis, as shown in Figure 3.4.

(i) As for the antenna, it is assumed that it is a microwave detector, such as 1N23B or the like, which is a microdipole antenna serving as a detector.

Now, the total electric field E simultaneously received by the dipole antenna for a standing-wave detection, which consists of the incident electric field intensity (E_1') by the transmitting antenna, the electric field intensities of the scattered wave $(E_{s1}'', E_{s2}'', E_{s3}'', \ldots)$ from the sample absorber, and the reflected electric field (E_2', E_3') from the floor and the wall surface, is determined by the following equation:

$$E = E_1' + E_{s1}'' + E_{s2}'' + E_{s3}'' + \cdots + E_2' + E_3' + \cdots. \tag{3.46}$$

The dipole antenna for standing-wave detection is installed parallel to the x axis and moves along the y axis to detect the standing wave. For this reason, we assume that the dipole antenna should be omnidirectional in the yz plane. As a concrete example of the total electric field E of Eq. (3.46),

$$
\begin{aligned}
E =& E_1' + E_{s1}'' + E_{s2}'' + E_{s3}'' + E_2' + E_3' \\
=& |E_1| e^{j\varphi_1} e^{\gamma_{n1} y} \\
& + |E_1| e^{j\varphi_1} (|\Gamma_{s1}| e^{j\varphi_{\Gamma s1}} e^{-\gamma_{ns1} y} + |\Gamma_{s2}| e^{j\varphi_{\Gamma s2}} e^{-\gamma_{ns2} y} + |\Gamma_{s3}| e^{j\varphi_{\Gamma s3}} e^{-\gamma_{ns3} y} + \cdots) \\
& + |E_2| e^{j\varphi_2} e^{-\gamma_{n2} y} + |E_3| e^{j\varphi_3} e^{\gamma_{n3} y}.
\end{aligned}
\tag{3.47}
$$

This is an example in which the total electric field E includes the electric field (E_1') directly incident on the dipole antenna and the three terms of the scattered electric field $(E_{s1}'', E_{s2}'', E_{s3}'')$ from the sample and two terms of the reflected electric field (E_2', E_3') from the wall and the floor. Note that the

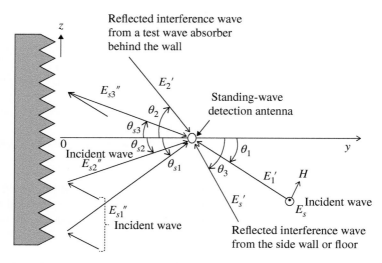

Figure 3.4 Example of a measurement model diagram in the case where there exist reflected waves and scattered waves from a finite length EM-wave absorber or other objects (in case of TE wave incidence.)

condition $|E''_{s1}| > |E''_{s2}| > |E''_{s3}| > \cdots$ is imposed onto the electric field intensity of the scattered waves from the absorber sample.

Here, $\varphi_k(k = 1, 2, 3, \ldots)$ is a phase angle that depends on directivity and $\varphi_{\Gamma s1}, \varphi_{\Gamma s2}, \ldots$, are phase angles related to the scattering coefficient $\Gamma_{sm}(m = 1, 2, 3, \ldots)$ of the sample absorber, where

$$\gamma_{nh} = \alpha_{nh} + j\beta_{nh} = (\alpha + j\beta)\cos\theta_h \quad (h = 1, 2, 3, \ldots)$$
$$\gamma_{nsi} = \alpha_{nsm} + j\beta_{nsi} = (\alpha + j\beta)\cos\theta_{si} \quad (i = 1, 2, 3, \ldots), \tag{3.48}$$

where γ_{nh} in Eq. (3.48) is the propagation constant associated with the incident waves of the transmission antenna and the reflected waves from the floor and the wall surface.

Also, γ_{nsi} represents the propagation constant of the scattered waves from the sample EM-wave absorber. The scattered waves and reflected waves from pyramidal absorbers, floors, etc., are assumed to be in such a state that only these y axis components propagate in the positive direction of the y axis. If the detector attached to the standing-wave detection antenna (dipole antenna) has square-law detection characteristics, the detection output voltage and current take a value proportional to $|E|^2$. Therefore, in order to know the standing-wave characteristics, $|E|^2$ in Eq. (3.47) is necessary.

In the following description, for simplicity of explanation, only the maximum intensity (E''_{s1}) among the scattered electric field $(E''_{s1}, E''_{s2}, E''_{s3}, \ldots)$ is assumed.

As a result, E in Eq. (3.47) is expressed simply by the following equation:

$$
\begin{aligned}
E =&|E_1|e^{j\varphi_1}e^{\gamma_{n1}y} + |E_1|e^{j\varphi_1}|\Gamma|e^{j\varphi_\Gamma}e^{-\gamma_{n1}y} \\
&+ |E_2|e^{j\varphi_2}e^{-\gamma_{n2}y} + |E_3|e^{j\varphi_3}e^{\lambda_{n3}y}.
\end{aligned}
\tag{3.49}
$$

In Eq. (3.49),

(a) *The first term on the right side*: Transmission incident wave E_1' is directly incident on the standing-wave detection antenna.

(b) *The second term on the right side*: The electric field E_{s1}'' with the maximum intensity among the scattered waves from the test sample EM-wave absorber and Γ: Scattering coefficient of EM-wave absorber as a sample.

(c) *The third term on the right side*: Reflected wave E_2' from the front or rear wall surface.

(d) *The fourth term on the right hand side*: Reflected wave E_3' from the floor or side wall.

When $|E|^2$ in Eq. (3.49) is calculated to know the detection output voltage (or current),

$$
\begin{aligned}
|E|^2 =&|E_1|^2 e^{2\alpha_{n1}y}(1 + |\Gamma|^2 e^{-4\alpha_{n1}y}) + |E_2|^2 e^{-2\alpha_{n2}y} + |E_3|^2 e^{2\alpha_{n3}y} && \text{①}\\
&+ 2|\Gamma||E_1{}^2|\cos(2\beta_{n1}y - \varphi_\Gamma) && \text{②}\\
&+ 2|E_1||E_2|[[|\Gamma|e^{-(\alpha_{n1}+\alpha_{n2})y}\cos\{(\beta_{n1} - \beta_{n2})y - (\varphi_1 - \varphi_2) - \varphi_\Gamma\} && \text{③}\\
&+ e^{(\alpha_{n1}-\alpha_{n2})y}\cos\{(\beta_{n1} + \beta_{n2})y + (\varphi_1 - \varphi_2)\}] && \text{④}\\
&+ 2|E_1||E_3|[e^{(\alpha_{n1}+\alpha_{n3})y}\cos\{(\beta_{n1} - \beta_{n3})y + (\varphi_1 - \varphi_3)\} && \text{⑤}\\
&+ 2|\Gamma|e^{(\alpha_{n3}-\alpha_{n1})}\cos\{(\beta_{n1} + \beta_{n3})y - (\varphi_1 - \varphi_3) - \varphi_\Gamma\} && \text{⑥}\\
&+ 2|E_2||E_3|e^{(\alpha_{n3}-\alpha_{n2})}\cos\{(\beta_{n2} + \beta_{n3})y - (\varphi_2 - \varphi_3)\} && \text{⑦},
\end{aligned}
\tag{3.50}
$$

where $\Gamma = |\Gamma|e^{j\varphi_\Gamma} = \dfrac{\eta_1-\eta_0}{\eta_1+\eta_0}$, $\eta_0 = \sqrt{\mu_0/\varepsilon_0}\cdot \sec\theta_1$, and η_1 is the input impedance of the test sample.

When representing the upper and lower envelope values at $y = 0$ as $|E_0|$ *max*, $|E_0|$*min* in front of the present test sample, respectively, the standing-wave ratio ρ can be fundamentally given by the following equation:

$$
\rho = \frac{|E_0|_{max}}{|E_0|_{min}} = \frac{1+|\Gamma|}{1-|\Gamma|}.
\tag{3.51}
$$

But in practice, the standing-wave shape changes generally depending on the behavior of each term from ① to ⑦ in Eq. (3.50). In the present case, the standing wave is generally classified into cases of generating a beat, the case of producing a swell, and the case composed of beat and swell.

In the following, let us consider the method of calculating a standing wave when a beat occurs [7].

3.4.1 Standing Wave Ratio in Beat Generation

In the present example, the beat generation is based on each of the terms ④, ⑥, and ⑦ in Eq. (3.50). In this case, although the term of ⑥ contains (Γ) which represents the scattering characteristics of the wave absorber, this term may be neglected since this term effect is usually very small (approximately 0.005–0.05). Further, the term ⑦ includes the product term of $|E_2| \cdot |E_3|$, but this term contribution is also small. (From experimental results, this term is evaluated as $|E_2|/|E_1| \leq |\Gamma|$, $|E_3|/|E_1| \leq |\Gamma|$.) For this reason, we think that the occurrence of beat is mainly due to ④.

After all, as the expression of $|E|^2$ in Eq. (3.50), we can finally adopt the expression adding ② and ④ to the expression excluding the right-side third term of the expression ① in Eq. (3.50). Figure 3.5 shows the state of beat generation drawn under these conditions.

Now, the standing wave ratio ρ is generally determined by the following equation

$$\rho = \sqrt{\frac{D + \frac{c}{2}}{D - \frac{c}{2}}}. \tag{3.52}$$

D represents the center value between the upper and lower average envelopes between points a and b when $y = 0$ as shown in Fig. 3.5. Regarding the

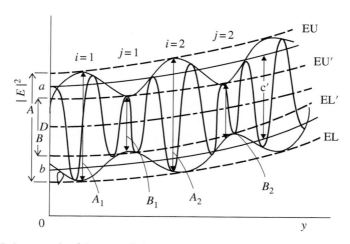

Figure 3.5 An example of the case of a beat.

determination of the value of c, the following two ways of thinking have been considered [7].

First, when the envelope of the standing wave is expressed by EU, EU′, EL, EL′ in the same figure, the following relations are held:

$$A = EU - EL \qquad (3.53)$$

$$B = EU' - EL'. \qquad (3.54)$$

The center line a and b in the upper and lower envelope of the standing wave,

$$a = \frac{1}{2}(EU + EU') \qquad (3.55)$$

$$b = \frac{1}{2}(EL' + EL). \qquad (3.56)$$

In addition, the width between a and b is

$$c' = a - b = 4|\Gamma||E_1|^2. \qquad (3.57)$$

The center line value D of the standing wave in the present case at $y = 0$ is defined as following expression, considering only the term having the large reflected field strength:

$$D = \left.\frac{EU + EL}{2}\right|_{y=0} = |E_1|^2(1 + |\Gamma|)^2 + |E_2|^2. \qquad (3.58)$$

In place of c in Eq. (3.52), by applying c' in Eq. (3.57) and D in Eq. (3.58) to Eq. (3.52), it become possible to calculate the present reflection coefficient.

(Note: The derivation processes of Eqs. (3.57) and (3.58) are given in the Appendix 3.A taking the example of scattering fields composed of four terms.)

In the present case, the standing-wave ratio on the front surface of the sample can be obtained from the beat drawing such as in Figure 3.5. However, this method is prone to errors when plotting the beat. Therefore, the following is considered as the method of calculating the standing-wave ratio [7]. This method defines the standing wave using the following c, paying attention to the maximum and minimum amplitudes A_i, B_j of the beat instead of c' of Eq. (3.57).

$$c = \frac{1}{2l}\left(\sum_{i=1}^{l} A_i + \sum_{j=1}^{l} B_j\right). \qquad (3.59)$$

Here, i and j are serial numbers counted from the front of the sample for the bellies and nodes of the beat wave, respectively, as shown in Figure 3.5.

It is now assumed that the bellies and nodes of standing-wave beats use the same numbers of the bellies and nodes next to each other Consequently, the standing wave ratio in this method can be given by Eq. (3.52).

This gives the standing-wave ratio on the EM-wave absorber sample surface when the beat is generated by the TE wave.

3.A Appendix

3.A.1 Appendix to Section 3.4.1 (1)

Explanation of the calculation method of Eq. (3.57) and (3.58)

In Eq. (3.47), the scattered wave from the test sample absorber (pyramidal EM-wave absorber) is treated here as four waves. The other reflected waves are treated in the same way as the case in Eq. (3.47).

That is, let us consider the case where the scattered waves from the test sample absorber are composed of four waves $(E_{s1}'', E_{s2}'', E_{s3}'', E_{s4}'')$, the electric field directly entering from the transmitting antenna is (E_1'), and the electric fields from the side wall or the floor composed of two waves (E_2', E_3'). The total electric field E in this case is

$$
\begin{aligned}
E =&|E_1|e^{j\varphi_1}e^{\gamma_{n1}y} + |E_1|e^{j\varphi_1}(|\Gamma_{s1}|e^{j\varphi_{\Gamma s1}}e^{-\gamma_{ns1}y} + |\Gamma_{s2}|e^{j\varphi_{\Gamma s2}}e^{-\gamma_{\Gamma s2}y} \\
&+ |\Gamma_{s3}|e^{j\varphi_{\Gamma s3}}e^{-\gamma_{ns3}y} + |\Gamma_{s4}|4e^{j\varphi_{\Gamma s4}}e^{-\gamma_{ns4}y}) + |E_2|e^{j\varphi_2}e^{-\gamma_{n2}y} + |E_3|e^{j\varphi_3}e^{\gamma_{n3}y}.
\end{aligned}
$$
(3.A.1)

A part of the square of equation in (3.A.1) is expressed as follows.

In the present case, only the main terms that generate beat and undulation are described, and the terms that have a slight effect on the standing-wave configuration are omitted in the following expression. Note that exponential function is represented by exp expression in order to make subscript of exponential function e easy to see.

$$
\begin{aligned}
|E|^2 =&|E_1|^2\{\exp(2\alpha_{n1}y) + |\Gamma_{s1}|^2\exp(-2\alpha_{n1}y) + |\Gamma_{s2}|^2\exp(-2\alpha_{ns2}y) \\
&+ |\Gamma_{s3}|^2\exp(-2\alpha_{ns3}y) + |\Gamma_{s4}|^2\exp(-2\alpha_{ns4}y)\} \qquad\qquad ① \\
&+ |E_2|^2\exp(-2\alpha_{n2}y) \\
&+ |E_3|^2\exp(2\alpha_{n3}y) \\
&+ 2|E_1|^2|\Gamma_{s1}|\exp\{(\alpha_{n1} - \alpha_{ns1})y\}\cos\{(\beta_{n1} + \beta_{ns1})y - \varphi_{\Gamma s1}\} \qquad ② \\
&+ 2|E_1|^2|\Gamma_{s2}|\exp\{(\alpha_{n1} - \alpha_{ns2})y\}\cos\{(\beta_{n1} + \beta_{ns2})y - \varphi_{\Gamma s2}\} \\
&+ \cdots
\end{aligned}
$$
(3.A.2)

As shown in Figure 3.5, the envelope of the standing wave is represented by EU, EU', EL', EL.

$$
\begin{aligned}
EU =&H_1 + H_2 + H_3 \\
EU' =&H_1 + H_2 - H_3 \\
EL' =&H_1 - H_2 + H_3 \\
EL =&H_1 - H_2 - H_3.
\end{aligned}
$$
(3.A.3)

Here, H_1, H_2, H_3 are expressed by the following equation in consideration of the strength of the reflected and scattered waves:

$$H_1 = |E_1|^2 \{\exp(2\alpha_{n1}y) + |\Gamma_{s1}|^2 \exp(-2\alpha_{n1}y) + |\Gamma_{s2}|^2 \exp(-2\alpha_{ns2}y)$$
$$+ |\Gamma_{s3}|^2 \exp(-2\alpha_{ns3}y) + |\Gamma_{s4}|^2 \exp(-2\alpha_{ns4}y)\}$$
$$+ |E_2|^2 \exp(-2\alpha_{n2}y) + |E_3|^2 \exp(2\alpha_{n3}y) \tag{3.A.4}$$

$$H_2 = 2|E_1|^2|\Gamma_{s1}| \exp(\alpha_{n1} - \alpha_{ns1})y \tag{3.A.5}$$

$$H_3 < 2|E_1|^2|\Gamma_{s2}|e^{(\alpha_{n1} - \alpha_{ns2})}$$
$$2|E_1|^2 + |\Gamma_{s3}|e^{(\alpha_{n1} - \alpha_{ns3})} \cdots . \tag{3.A.6}$$

Here, H_1 is a term of ① in Eq. (3.A.2) and H_2 is an amplitude term of ② in Eq. (3.A.2).

Referring to Figure 3.5, the upper and lower centerlines of the standing wave,

$$a = \frac{1}{2}(\text{EU} + \text{EU}')$$
$$= (H_1 + H_2)$$
$$= |E_1|^2 \{\exp(2\alpha_{n1}y) + |\Gamma_{s1}|^2 \exp(-2\alpha_{n1}y) + |\Gamma_{s2}|^2 \exp(-2\alpha_{ns2}y)$$
$$+ |\Gamma_{s3}|^2 \exp(-2\alpha_{ns3}y) + |\Gamma_{s4}|^2 \exp(-2\alpha_{ns4}y)\}$$
$$+ |E_2|^2 \exp(-2\alpha_{n2}y) + |E_3|^2 \exp(2\alpha_{n3}y) + 2|E_1|^2|\Gamma_{s1}| \exp(\alpha_{n1} - \alpha_{ns1})y \tag{3.A.7}$$

$$b = \frac{1}{2}(\text{EL}' + \text{EL})$$
$$= (H_1 - H_2)$$
$$= |E_1|^2 \{\exp(2\alpha_{n1}y) + |\Gamma_{s1}|^2 \exp(-2\alpha_{n1}y) + |\Gamma_{s2}|^2 \exp(-2\alpha_{ns2}y)$$
$$+ |\Gamma_{s3}|^2 \exp(-2\alpha_{ns3}y) + |\Gamma_{s4}|^2 \exp(-2\alpha_{ns4}y)\}$$
$$+ |E_2|^2 \exp(-2\alpha_{n2}y) + |E_3|^2 \exp(2\alpha_{n3}y) - 2|E_1|^2|\Gamma_{s1}| \exp(\alpha_{n1} - \alpha_{ns1})y. \tag{3.A.8}$$

As a result, Eq. (3.57) is

$$c' = (a - b)|_{y=0} = 4|E_1|^2|\Gamma_{s1}|. \tag{3.57}$$

Further, the value of D at $y = 0$ of the center line of the standing wave,

$$D = \frac{\text{EU} + \text{EL}}{2}\bigg|_{y=0} = H_1|_{y=0}$$
$$= |E_1|^2(1 + |\Gamma_{s1}|^2 + |\Gamma_{s2}|^2 + |\Gamma_{s3}|^2 + |\Gamma_{s4}|^2)$$
$$+ |E_2|^2 + |E_3|^2. \tag{3.A.9}$$

Here, when only the electric field (E_1') from the transmitting antenna, the maximum scattered field (E_{s1}'') from the sample absorber, and the reflected electric field (E_2') from the walls located in front of and behind the sample surface as the electric fields, exist, as in the case in Section 3.4.1, the relation of D can be represented in the following equation:

$$D = \frac{EU + EL}{2}\bigg|_{y=0} = |E_1|^2(1 + |\Gamma|)^2 + |E_2|^2. \tag{3.58}$$

References

1 Dewitt, B.T. and Burnsid, W.D. (1988). Electromagnetic scattering by pyramidal and wedge absorber. *IEEE Trans. Antennas Propag.* 36 (7): 971–984.

2 Vandrer Vorst, A., Rosen, A., and Kotsuka, Y. (2006). *RF/Microwave Interaction with Biological Tissue.* Wiley Interscience.

3 Yang, C.F., Burnside, W.D., and Rudduck, R.C. (1992). A periodic moment method solution for TM scattering from lossy dielectric bodies with application to wedge absorber. *IEEE Trans. Antennas Propag.* 40 (9): 652–660.

4 Yang, C.F., Burnside, W., and Rudduck, R.C. (1993). A doubly periodic moment method solution for the analysis and design of an absorber covered wall. *IEEE Trans. Antennas Propag.* 41 (5): 600–609.

5 Sun, W., Liu, K., and Balanis, C.A. (1996). Analysis of singly and doubly periodic absorbers by frequency-domain finite-difference method. *IEEE Trans. Antennas Propag.* 44 (6): 798–805.

6 Shimizu, Y. (Editorial Committee Chairman) (1999). Electromagnetic waves absorption and shielding. Published by Nikkei Gijyutu Tosho in Japan, p. 123, 1999.

7 Ono, M., Shibuya, T., and Hsieh, K.-C. (1989). A practical method of measuring scattering characteristics a pyramidal absorber. *IEEE Trans. Electromagn. Compat.* 31 (3): 312–316.

4

Basic Theory of Computer Analysis

One of the reasons that a wide variety of EM-wave absorber analyses came to be possible recently is the remarkable progress of modeling and computer simulation technologies. In addition, in actual computer simulation technologies, a number of theoretical analysis methods and means are compounded and systemized. In this chapter, in order to deepen the basic understanding of computer simulation analysis of EM-wave absorbers, the finite-difference time-domain (FDTD) method and finite element (FE) method are explained in detail, including their fundamental principles. In Section 4.1, first, after the history of the FDTD method is briefly described, the analytical procedures of the FDTD method are described in detail, step by step. In particular, the idea that forms the basis of the FDTD analysis method is presented in detail.

First, by taking up Maxwell's equations without the magnetic current source, how to discretize the electromagnetic fields in the space–time region in the case of FDTD analysis is explained. In addition, the detailed FDTD analysis processes are provided in the case of Maxwell's general equations when including the current source and magnetic current source. As a concrete example of numerical calculation, taking an example of a ferrite absorber with holes, the methods of setting the periodic boundary conditions, variable cell size problem, convergence problem of analytical solution against cell numbers, and the like, are demonstrated.

Section 4.2 includes detailed descriptions of the FE method.

First, the variational method that forms the way of thinking of the FE method is explained, taking the optical path problem following Fermat's principle as an example.

Next, a general description is clarified in a form that lists the analytical procedures of the FE method. As concrete analysis methods, in Section 4.2.3, an example of the two-dimensional electrostatic field analysis is explained on the basis of the variational principle in the case where the functional is known. Further, in Section 4.3, as a concrete analysis method when the functional is unknown, the three-dimensional current vector potential method based on the Galerkin method, called a direct method in the variational method, is described

Electromagnetic Wave Absorbers: Detailed Theories and Applications, First Edition. Youji Kotsuka.
© 2019 John Wiley & Sons, Inc. Published 2019 by John Wiley & Sons, Inc.

in detail, and includes the derivation methods of the expressions. In this case, the discretization method for numerical analysis of basic equations and auxiliary equations is particularly explained in detail. Finally, the eddy current distributions at the time of inductive heating and an eddy current absorber to absorb unnecessary eddy current are introduced as an analysis example of the three-dimensional current vector potential method.

4.1 FDTD Analysis Method

4.1.1 Basis of FDTD

If stated simply, FDTD analysis is a method for solving electromagnetic-field problems which are based on Maxwell's equations, by applying the calculus of finite differences to each component of electromagnetic fields and the time term. Usually, this method is referred to as "finite-difference time-domain method." This has been widely used as a powerful analytical tool in the fields of electromagnetic-wave studies.

The basic algorithm of the FDTD method was proposed by K. S. Yee in 1966 [1].

After that, the FDTD method has been studied by many researchers. A. Taflove attempted to apply it to various electromagnetic-wave problems [2]. From the boundary conditions viewpoint, G. Mur [3] proposed a method of absorbing boundary, and J. P. Berenger [4] also proposed a perfectly matched layer (PML) absorption boundary which plays an important role in analyzing the electromagnetic-wave problems in a closed region. In this way, the FDTD method was gradually established as a powerful tool for the electromagnetic-field analysis.

Basic points of the concept of the FDTD method are summarized as follows:

(a) First, define the analysis region, including areas such as the wave source and the scattering body.
(b) Both the arrangements of the electromagnetic-field component and time are represented by dividing into very small cells in the analysis region.
(c) When representing a three-dimensional unit cell size using Δx, Δy, and Δz, the numbers (i, j, k) are arranged at lattice points that are vertices in a small divided rectangular body.

The electric and magnetic fields are arranged alternately in time. In our case, the electric field is represented using an integer order, and the magnetic field is represented by a semi-odd order.
(d) In this difference method, we introduce the spatiotemporal analysis method. Although there exist the central difference, forward difference, and backward difference methods in the calculus of finite differences, the central difference method, as shown in Figure 4.1, is usually adopted for its accuracy.

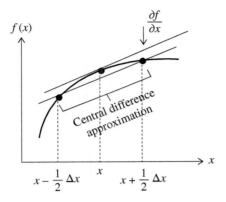

Figure 4.1 Central difference.

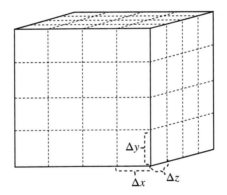

Figure 4.2 Closed analysis area.

(e) In this analysis, pay attention to the method of time axis arrangement to each of the electromagnetic fields and their notation methods.

The FDTD analysis method is basically a method of analyzing a closed area in an analysis region, consisting of a microscopic rectangular parallelepiped, as shown in Figure 4.2. Therefore, when dealing with a problem such as the EM-wave absorber being arranged in a space extending infinitely, the analysis method must be converted into a closed region using the absorption boundary condition and the periodic boundary condition. In the following, the procedures for determining the analytical region, so as to be able to surround the wave source and the scattering body, are also explained. Finally, we derive the difference equation which is the FDTD expression based on the method called Yee's algorithm [1].

4.1.2 Methods of Time and Space Difference

As is well known, the component expressions of Maxwell's equations consist of the electromagnetic-field components and time terms. When generally representing these electromagnetic-field components using the notation f, which suggests a function, each central difference in time and space can be expressed as the following equation. In the difference method, the central difference is usually adopted from the viewpoint of accuracy as was pointed out earlier.

$$\frac{\partial f}{\partial x} \approx \frac{f\left(x + \frac{\Delta x}{2}, y, z, t\right) - f\left(x - \frac{\Delta x}{2}, y, z, t\right)}{\Delta x} \tag{4.1}$$

$$\frac{\partial f}{\partial t} \approx \frac{f\left(x, y, z, t + \frac{\Delta t}{2}\right) - f\left(x, y, z, t - \frac{\Delta t}{2}\right)}{\Delta t}. \tag{4.2}$$

In this case, f corresponds to each component of the electric or magnetic fields.

Here, Δt means a discrete time interval, as explained later in Figure 4.3.

The analytical region is now represented by finely divided cells, as suggested in Figure 4.2. Also, since it is necessary to discretize time in the FDTD analysis, when a point at the vertex of each rectangular parallelepiped is expressed by (x, y, z, t), this expression is usually rewritten using integers (i, j, k, n) as follows:

$$(x, y, z, t) = (i\Delta x, j\Delta y, k\Delta z, n\Delta t). \tag{4.3}$$

Here, n means an integer number, and each designation $\Delta x, \Delta y,$ and Δz denote the length of the rectangular sides of the unit cell, and they are called the cell size. Δt is called a time step.

Accordingly, the essentially continuous variables (x, y, z, t) are discretized as described.

Furthermore, in FDTD analysis, the notations of electromagnetic fields and time are simplified, as follows after omitting the $\Delta x, \Delta y, \Delta z,$ and Δt expressions and using only (i, j, k, n). For this reason, when Expression (4.3) is expressed by the functional form of f, it is expressed as follows:

$$f(x, y, z, n) = f^n(i, j, k). \tag{4.4}$$

Therefore, (i, j, k) is in equivalent relationship with the coordinates on each grid point.

By means of introducing these methods, Eqs. (4.1) and (4.2) are finally rewritten as follows:

$$\frac{\partial f}{\partial x} \approx \frac{f^n\left(i + \frac{1}{2}, j, k\right) - f^{n-1}\left(i - \frac{1}{2}, j, k\right)}{\Delta x} \tag{4.5}$$

$$\frac{\partial f}{\partial t} \approx \frac{f^{n+\frac{1}{2}}(i,j,k) - f^{n-\frac{1}{2}} - (i,j,k)}{\Delta t}. \tag{4.6}$$

4.1.3 Relationship of Time Arrangement of the Electromagnetic Field

As the FDTD analysis was described in the previous section, each of the electric and magnetic fields is arranged alternately in time, because the central difference method is adopted in the present case.

In this case, an electric field can be expressed as the time of an integer order and a magnetic field can be expressed as the time of a half-odd order, and, of course, their reverses can be considered. We can choose either of them.

In Figure 4.3, the case of electric fields is assigned to each integer order time of $t = (n - 1)\Delta t, n\Delta t, (n + 1)\Delta t$ and the case of magnetic fields is assigned to each time with a half-odd order of $t = (n - 1/2)\Delta t, (n + 1/2)\Delta t, \dots$, respectively.

Consequently, in the actual calculation, E^n can be determined from the electric field E^{n-1} at $t = (n - 1)\Delta t$ and the magnetic field $H^{n-\frac{1}{2}}$ at $t = (n - 1/2)\Delta t$.

Further, $H^{n+\frac{1}{2}}$ is determined from $H^{n-\frac{1}{2}}$ and E^n. In this manner, the electric and magnetic fields are calculated along with the time axis. In order to concretely determine the difference expression, lets us consider the following Maxwell's equations which omit the magnetic current term for simplicity:

$$\nabla \times H = \varepsilon \frac{\partial E}{\partial t} + \sigma E \tag{4.7}$$

$$\nabla \times E = -\mu \frac{\partial H}{\partial t}. \tag{4.8}$$

Here, μ is magnetic permeability, σ is the conductivity, and ε is permittivity.

When we represent Eqs. (4.7) and (4.8) using the time term,

$$\frac{\partial E}{\partial t} = -\frac{\sigma}{\varepsilon} E + \frac{1}{\varepsilon} \nabla \times H \tag{4.9}$$

$$\frac{\partial H}{\partial t} = -\frac{1}{\mu} \nabla \times E. \tag{4.10}$$

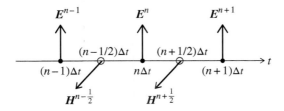

Figure 4.3 Electromagnetic field arrangements on the time axis.

According to Figure 4.3, since the times of the present electric fields are located in $t = (n - 1)\Delta t$ and $n\Delta t$, the time derivative of the electric field on the left side in Eq. (4.9) has to be carried out by $t = (n - 1/2)\Delta t$ at the middle time point.[1]

Also, since the times of the magnetic fields are located at $t = (n - 1/2)\Delta t$ and $(n + 1/2)\Delta t$, the time derivative of the magnetic field on the left side in Eq. (4.10) should be performed in the middle point $t = n\Delta t$ between these points.

According to the notation (4.5) and (4.6) in Section 4.1.2, Eqs. (4.9) and (4.10) are represented by the following equations, respectively:

$$\left.\frac{\partial E}{\partial t}\right|_{t = (n-1/2)\Delta t} = \frac{E^n - E^{n-1}}{\Delta t} \tag{4.11}$$

$$\left.\frac{\partial H}{\partial t}\right|_{t = n\Delta t} = \frac{H^{n+\frac{1}{2}} - H^{n-\frac{1}{2}}}{\Delta t}. \tag{4.12}$$

Substituting Eqs. (4.11) and (4.12) into Eqs. (4.9) and (4.10), respectively,

$$\frac{E^n - E^{n-1}}{\Delta t} = -\frac{\sigma}{\varepsilon}E^{n-\frac{1}{2}} + \frac{1}{\varepsilon}\nabla \times H^{n-\frac{1}{2}} \tag{4.13}$$

$$\frac{H^{n+\frac{1}{2}} - H^{n-\frac{1}{2}}}{\Delta t} = -\frac{1}{\mu}\nabla \times E^n. \tag{4.14}$$

Here, we should note that the electric field of the first term on the right-hand side of Eq. (4.13) is expressed at the time $t = (n - 1/2)\Delta t$ at which the magnetic field should be taken.

As is apparent from Figure 4.3, the electric field must be represented at the time $t = (n - 1)\Delta t$ or $n\Delta t$. By means of the countermeasure of modifying this electric field expression, the following approximation that has been evaluated as a highly accurate method is employed:

$$\sigma E^{n-\frac{1}{2}} \approx \sigma \frac{E^{n-1} + E^n}{2}. \tag{4.15}$$

When Eq. (4.13) is rewritten using Eq. (4.15),

$$\frac{E^n - E^{n-1}}{\Delta t} = -\frac{\sigma}{\varepsilon}\frac{E^{n-1} - E^n}{2} + \frac{1}{\varepsilon}\nabla \times H^{n-\frac{1}{2}}. \tag{4.16}$$

Solving this equation for E^n,

$$E^n = \frac{1 - \sigma\Delta t/2\varepsilon}{1 + \sigma\Delta t/2\varepsilon}E^{n-1} + \frac{\Delta t/\varepsilon}{1 + \sigma\Delta t/2\varepsilon}\nabla \times H^{n-\frac{1}{2}}. \tag{4.17}$$

---

1 In this case, in Figure 4.3, an expression related to E^{n+1} by differentiating at the time of $t = (n + 1/2)\Delta t$ can also be considered. However, we here adopt the electric field of time differentiating at the point $t = (n - 1/2)\Delta t$, because we intend so as to be able to use the electric field as an initial condition or an incident wave source.

That is, from the electric field E^{n-1} of the time $t = (n-1)\Delta t$ and the magnetic fields $H^{n-\frac{1}{2}}$ of the time $t = (n-1/2)\Delta t$, the electric field E^n after a next half step can be calculated.

Further, referring to Figure 4.3, since the magnetic field is calculated after a half step of the electric field E^n, the magnetic field $H^{n+\frac{1}{2}}$ can be determined from Eq. (4.14).

$$H^{n+\frac{1}{2}} = H^{n-\frac{1}{2}} - \frac{\Delta t}{\mu} \nabla \times E^n. \tag{4.18}$$

4.1.4 Relationship of Spatial Arrangement of the Electromagnetic Field

In the FDTD analysis described here, the central difference, as shown in Figure 4.1, is employed. Therefore, in the space coordinate system also, the electric field and the magnetic field are alternately arranged according to this central difference.

In this case, the electric fields in the unit cells are arranged along each side of the unit cells, as shown in Figure 4.4. On the other hand, the magnetic fields are assigned perpendicular to the center of the unit cell surfaces. The reasons for

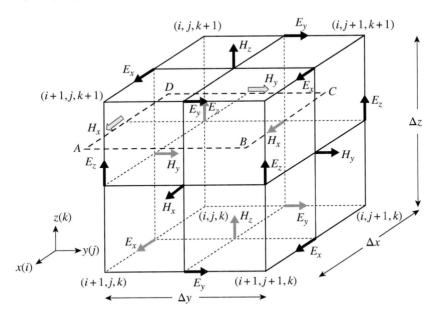

Figure 4.4 Spatial arrangement of the electromagnetic fields when focusing on one unit cell in the whole analysis area. (The magnetic fields on each side of the square ABCD are also depicted.)

these electromagnetic-field arrangements are clear from the principles of the Maxwell equation described in Section 2.3.1.

Taking here the z-component E_z of the electric field as an example, let us derive the FDTD expression in the form of Eq. (4.4). From Eq. (4.17), the component of the electric field is first expressed by the following equation:

$$E_z^n = \frac{1 - \sigma \Delta t / 2\varepsilon}{1 + \sigma \Delta t / 2\varepsilon} E_z^{n-1} + \frac{\Delta t / \varepsilon}{1 + \sigma \Delta t / 2\varepsilon} \left(\frac{\partial H_y^{n-\frac{1}{2}}}{\partial x} - \frac{\partial H_x^{n-\frac{1}{2}}}{\partial y} \right). \tag{4.19}$$

In this case, E_z is assigned to the four positions of the unit cells shown in Figure 4.4.

The choice of E_z is clear from the principles of the FDTD method, which is governed by the time arrangements in Figure 4.3. That is, the electric field $E_z(i, j, k + 1/2)$ as shown in Figures 4.4 and 4.5, which is at the shortest distance from the reference point of the spatial arrangement of the unit cells (i, j, k), should be selected. Figure 4.5 also shows the positions of the magnetic fields surrounding a center electric field of $E_z(i, j, k + 1/2)$. On the other hand, referring to Figures 4.4 and 4.5, the differential field representation of the magnetic field in the second term of the left-hand side of Expression (4.19) is

$$\left. \frac{\partial H_y^{n-\frac{1}{2}}}{\partial x} \right|_{\left(i,j,k+\frac{1}{2} \right)} = \frac{H_y^{n-\frac{1}{2}} \left(i + \frac{1}{2}, j, k + \frac{1}{2} \right) - H_y^{n-\frac{1}{2}} \left(i - \frac{1}{2}, j, k + \frac{1}{2} \right)}{\Delta x}$$

(4.20)

$$\left. \frac{\partial H_x^{n-\frac{1}{2}}}{\partial y} \right|_{\left(i,j,k+\frac{1}{2} \right)} = \frac{H_x^{n-\frac{1}{2}} \left(i, j + \frac{1}{2}, k + \frac{1}{2} \right) - H_x^{n-\frac{1}{2}} \left(i, j - \frac{1}{2}, k + \frac{1}{2} \right)}{\Delta y}.$$

(4.21)

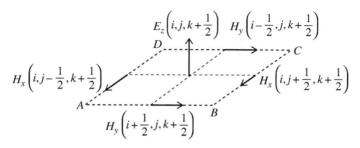

Figure 4.5 Arrangement of the magnetic field around the electric field at lattice point $(i, j, k + 1/2)$.

When Eqs. (4.20) and (4.21) are substituted into Eq. (4.19), the z-component of the electric field is expressed in the form of the following FDTD expression:

$$
\therefore E_x^n \left(i + \frac{1}{2}, j, k \right) = \frac{2\varepsilon \left(i + \frac{1}{2}, j, k \right) - \Delta t \sigma \left(i + \frac{1}{2}, j, k \right)}{2\varepsilon \left(i + \frac{1}{2}, j, k \right) + \Delta t \sigma \left(i + \frac{1}{2}, j, k \right)}
$$

$$
\times E_x^{n-1} \left(i + \frac{1}{2}, j, k \right) + \frac{2\Delta t}{2\varepsilon \left(i + \frac{1}{2}, j, k \right) + \Delta t \sigma \left(i + \frac{1}{2}, j, k \right)}
$$

$$
\left[\frac{H_z^{n-\frac{1}{2}} \left(i + \frac{1}{2}, j + \frac{1}{2}, k \right) - H_z^{n-\frac{1}{2}} \left(i + \frac{1}{2}, j - \frac{1}{2}, k \right)}{\Delta y} \right.
$$

$$
\left. + \frac{H_y^{n-\frac{1}{2}} \left(i + \frac{1}{2}, j, k - \frac{1}{2} \right) - H_y^{n-\frac{1}{2}} \left(i + \frac{1}{2}, j, k + \frac{1}{2} \right)}{\Delta z} \right]. \tag{4.22}
$$

In this case, because we are considering an isotropic medium, the spatial arrangement positions (i, j, k) on the material constants are omitted.

By the abovementioned procedures, the FDTD expressions of each electromagnetic-field component can be derived. In the following section, as the general expressions of the three-dimensional FDTD analysis, we consider the cases including both the magnetic current term and the electric current term in Maxwell's equation.

4.1.5 General Expressions of FDTD Analysis

In this section, from the general Maxwell's equations, including the electric current term and the magnetic current term, the methods for deriving the FDTD expressions are briefly introduced.

Methods for deriving FDTD expressions are basically the same as in Sections 4.1.3 and 4.1.4. First, the general Maxwell equations are represented by the following equations:

$$
\nabla \times H = \frac{\partial D}{\partial t} + J \tag{4.23}
$$

$$
\nabla \times E = -\frac{\partial B}{\partial t} - J^*. \tag{4.24}
$$

Here, the $E, H, D,$ and B represent the electric field, the magnetic field, the electric flux density, and the magnetic flux density, respectively.

The J and J^* represent the electric current density and the magnetic current density, respectively. By substituting the relationships $B = \mu H, D = \varepsilon E$,

$J = \sigma E$, and $J^{*} = \sigma^{*} H$ into the Maxwell's equations (4.23) and (4.24),

$$\nabla \times H = \varepsilon \frac{\partial E}{\partial t} + \sigma E \qquad (4.25)$$

$$\nabla \times E = -\mu \frac{\partial H}{\partial t} - \sigma^{*} H. \qquad (4.26)$$

In the symbols in these equations, σ and σ^{*} denote a conductivity and a magnetic permeability, respectively. Also, ε and μ are the dielectric constant and permeability, respectively.

If these expressions are rewritten in the component forms of three-dimensional orthogonal coordinates, respectively,

$$\left(\frac{\partial H_z}{\partial y} - \frac{\partial H_y}{\partial z} \right) = \varepsilon \frac{\partial E_x}{\partial t} + \sigma E_x \qquad (4.27)$$

$$\left(\frac{\partial H_x}{\partial z} - \frac{\partial H_z}{\partial x} \right) = \varepsilon \frac{\partial E_y}{\partial t} + \sigma E_y \qquad (4.28)$$

$$\left(\frac{\partial H_y}{\partial x} - \frac{\partial H_x}{\partial y} \right) = \varepsilon \frac{\partial H_z}{\partial t} + \sigma E_z \qquad (4.29)$$

$$\left(\frac{\partial E_z}{\partial y} - \frac{\partial E_y}{\partial z} \right) = -\mu \frac{\partial H_x}{\partial t} - \sigma^{*} H_x \qquad (4.30)$$

$$\left(\frac{\partial E_x}{\partial z} - \frac{\partial E_z}{\partial x} \right) = -\mu \frac{\partial H_y}{\partial t} - \sigma^{*} H_y \qquad (4.31)$$

$$\left(\frac{\partial E_y}{\partial x} - \frac{\partial E_x}{\partial y} \right) = -\mu \frac{\partial H_z}{\partial t} - \sigma^{*} H_z. \qquad (4.32)$$

Next, when the equations from (4.27) to (4.32) are expressed by the time derivative terms,

$$\frac{\partial E_x}{\partial t} = \frac{1}{\varepsilon} \left(\frac{\partial H_z}{\partial y} - \frac{\partial H_y}{\partial z} - \sigma E_x \right) \qquad (4.33)$$

$$\frac{\partial E_y}{\partial t} = \frac{1}{\varepsilon} \left(\frac{\partial H_x}{\partial z} - \frac{\partial H_z}{\partial x} - \sigma E_y \right) \qquad (4.34)$$

$$\frac{\partial E_z}{\partial t} = \frac{1}{\varepsilon} \left(\frac{\partial H_y}{\partial x} - \frac{\partial H_x}{\partial y} - \sigma E_z \right) \qquad (4.35)$$

$$\frac{\partial H_x}{\partial t} = \frac{1}{\mu} \left(\frac{\partial E_y}{\partial z} - \frac{\partial E_z}{\partial y} + \sigma^{*} H_x \right) \qquad (4.36)$$

$$\frac{\partial H_y}{\partial t} = \frac{1}{\mu} \left(\frac{\partial E_z}{\partial x} - \frac{\partial E_x}{\partial z} + \sigma^{*} H_y \right) \qquad (4.37)$$

$$\frac{\partial H_z}{\partial t} = \frac{1}{\mu} \left(\frac{\partial E_x}{\partial y} - \frac{\partial E_y}{\partial x} + \sigma^{*} H_z \right). \qquad (4.38)$$

Incidentally, when analyzing a lossy medium such as an EM-wave absorber, the permittivity and permeability are expressed by complex numbers. In addition, complex permittivity and complex permeability are often represented by frequency dispersion equations.

In this case, note that the dielectric constant and permeability of the formulas (4.33)–(4.38) are related to the values of σ, σ^* as follows:

$$\varepsilon = \varepsilon_0 \varepsilon_r' - j\frac{\sigma}{\omega}$$

$$\mu = \mu_0 \mu_r' - j\frac{\sigma^*}{\omega}.$$

Here, $\sigma = \omega\varepsilon_0\varepsilon_r''$, $\sigma^* = \omega\mu_0\mu_r''$,

where ε_0 is the permittivity in vacuum, ε_r'' is the imaginary part of the complex relative permittivity, μ_0 is the permeability in vacuum, and μ_r'' is the imaginary part of the complex relative permeability. ω is the angular frequency.

Using these transformation equations, it becomes possible to analyze the complex permittivity and the complex permeability problem in the FDTD method.

In the present case, as shown in Figure 4.3 in Section 4.1.3, the electric field is taken as the time reference and the discrete time interval between the electric field and the magnetic field is defined as the same value $\Delta t/2$. Therefore, also from the viewpoint of this time reference, the time notation in which the magnetic field is expressed by H^n or the electric field is represented using $E^{n-\frac{1}{2}}$ is not permitted, respectively. As pointed out in Eq. (4.13), the irrationality of these times should be treated by the following approximate expression:

$$H^n = \frac{H^{n+\frac{1}{2}} + H^{n-\frac{1}{2}}}{2} \tag{4.39}$$

$$\left.\frac{\partial E}{\partial t}\right|_{t=\left(n-\frac{1}{2}\right)\Delta t} = \frac{E^n - E^{n-1}}{\Delta t}. \tag{4.40}$$

Now, under these conditions, we can derive the equations relating to the FDTD method for Eqs. (4.33)–(4.35).

$$E_x^n\left(i+\frac{1}{2},j,k\right) = \frac{1 - \frac{\sigma\Delta t}{2\varepsilon}}{1 + \frac{\sigma\Delta t}{2\varepsilon}} E_x^{n-1}\left(i+\frac{1}{2},j,k\right)$$

$$+ \frac{\Delta t/\varepsilon}{1 + \frac{\sigma\Delta t}{2\varepsilon}} \frac{1}{\Delta y} \left\{ H_z^{n-\frac{1}{2}}\left(i+\frac{1}{2},j+\frac{1}{2},k\right) - H_z^{n-\frac{1}{2}}\left(i+\frac{1}{2},j-\frac{1}{2},k\right) \right\}$$

$$- \frac{\Delta t/\varepsilon}{1 + \frac{\sigma\Delta t}{2\varepsilon}} \frac{1}{\Delta z} \left\{ H_y^{n-\frac{1}{2}}\left(i+\frac{1}{2},j,k+\frac{1}{2}\right) - H_y^{n-\frac{1}{2}}\left(i+\frac{1}{2},j,k-\frac{1}{2}\right) \right\}$$

$$\tag{4.41}$$

$$E_y^n\left(i, j+\frac{1}{2}, k\right) = \frac{1 - \frac{\sigma \Delta t}{2\varepsilon}}{1 + \frac{\sigma \Delta t}{2\varepsilon}} E_y^{n-1}\left(i, j+\frac{1}{2}, k\right)$$

$$+ \frac{\Delta t/\varepsilon}{1 + \frac{\sigma \Delta t}{2\varepsilon}} \frac{1}{\Delta z}\left\{H_x^{n-\frac{1}{2}}\left(i, j+\frac{1}{2}, k+\frac{1}{2}\right) - H_x^{n-\frac{1}{2}}\left(i, j+\frac{1}{2}, k-\frac{1}{2}\right)\right\}$$

$$- \frac{\Delta t/\varepsilon}{1 + \frac{\sigma \Delta t}{2\varepsilon}} \frac{1}{\Delta x}\left\{H_z^{n-\frac{1}{2}}\left(i+\frac{1}{2}, j+\frac{1}{2}, k\right) - H_z^{n-\frac{1}{2}}\left(i-\frac{1}{2}, j+\frac{1}{2}, k\right)\right\}$$

$$\tag{4.42}$$

$$E_z^n\left(i, j, k+\frac{1}{2}\right) = \frac{1 - \frac{\sigma \Delta t}{2\varepsilon}}{1 + \frac{\sigma \Delta t}{2\varepsilon}} E_z^{n-1}\left(i, j, k+\frac{1}{2}\right)$$

$$+ \frac{\Delta t/\varepsilon}{1 + \frac{\sigma \Delta t}{2\varepsilon}} \frac{1}{\Delta x}\left\{H_y^{n-\frac{1}{2}}\left(i+\frac{1}{2}, j, k+\frac{1}{2}\right) - H_y^{n-\frac{1}{2}}\left(i-\frac{1}{2}, j, k+\frac{1}{2}\right)\right\}$$

$$- \frac{\Delta t/\varepsilon}{1 + \frac{\sigma \Delta t}{2\varepsilon}} \frac{1}{\Delta y}\left\{H_x^{n-\frac{1}{2}}\left(i, j+\frac{1}{2}, k+\frac{1}{2}\right) - H_x^{n-\frac{1}{2}}\left(i, j-\frac{1}{2}, k+\frac{1}{2}\right)\right\}.$$

$$\tag{4.43}$$

As described in Section 4.1.2, Δt is the time discrete interval and Δx, Δy, and Δz represent the cell size in the x, y, and z directions, respectively. i, j, and k represent integers indicating the position of the three-dimensional regions, and n expresses an integer number of the time steps. Looking at these equations, it can be seen that the electric field at time n on the left-hand side is composed of the electric field of $n-1$ on the right-hand side and the magnetic field of $n-1/2$. That is, these equations follow the provisions of the FDTD method.

Similarly, using the relations of Eqs. (4.36)–(4.38) and the following Eqs. (4.44) and (4.45), the time difference and space difference expressions for the magnetic field components can be given.

$$E^{n+\frac{1}{2}} = \frac{E^n + E^{n-1}}{2} \tag{4.44}$$

$$\left.\frac{\partial H}{\partial t}\right|_{t=n\Delta t} = \frac{H^{n+\frac{1}{2}} - H^{n-\frac{1}{2}}}{\Delta t} \tag{4.45}$$

$$H_x^{n+\frac{1}{2}}\left(i, j+\frac{1}{2}, k+\frac{1}{2}\right) = \frac{1 - \frac{\sigma^* \Delta t}{2\mu}}{1 + \frac{\sigma^* \Delta t}{2\mu}} H_x^{n-\frac{1}{2}}\left(i, j+\frac{1}{2}, k+\frac{1}{2}\right)$$

$$- \frac{\Delta t/\mu}{1 + \frac{\sigma^* \Delta t}{2\mu}} \frac{1}{\Delta y}\left\{E_z^n\left(i, j+1, k+\frac{1}{2}\right) - E_z^n\left(i, j, k+\frac{1}{2}\right)\right\}$$

$$+ \frac{\Delta t/\mu}{1 + \frac{\sigma^* \Delta t}{2\mu}} \frac{1}{\Delta z}\left\{E_y^n\left(i, j+\frac{1}{2}, k+1\right) - E_y^n\left(i, j+\frac{1}{2}, k\right)\right\} \tag{4.46}$$

$$H_y^{n+\frac{1}{2}}\left(i+\frac{1}{2},j,k+\frac{1}{2}\right) = \frac{1 - \frac{\sigma^*\Delta t}{2\mu}}{1 + \frac{\sigma^*\Delta t}{2\mu}} H_y^{n-\frac{1}{2}}\left(i+\frac{1}{2},j,k+\frac{1}{2}\right)$$

$$-\frac{\Delta t/\mu}{1+\frac{\sigma^*\Delta t}{2\mu}}\frac{1}{\Delta z}\left\{E_x^{n}\left(i+\frac{1}{2},j,k+1\right) - E_x^{n}\left(i+\frac{1}{2},j,k\right)\right\}$$

$$+\frac{\Delta t/\mu}{1+\frac{\sigma^*\Delta t}{2\mu}}\frac{1}{\Delta x}\left\{E_z^{n}\left(i+1,j,k+\frac{1}{2}\right) - E_z^{n}\left(i,j,k+\frac{1}{2}\right)\right\} \qquad (4.47)$$

$$H_z^{n+\frac{1}{2}}\left(i+\frac{1}{2},j+\frac{1}{2},k\right) = \frac{1 - \frac{\sigma^*\Delta t}{2\mu}}{1 + \frac{\sigma^*\Delta t}{2\mu}} H_z^{n-\frac{1}{2}}\left(i+\frac{1}{2},j+\frac{1}{2},k\right)$$

$$-\frac{\Delta t/\mu}{1+\frac{\sigma^*\Delta t}{2\mu}}\frac{1}{\Delta x}\left\{E_y^{n}\left(i+1,j+\frac{1}{2},k\right) - E_y^{n}\left(i,j+\frac{1}{2},k\right)\right\}$$

$$+\frac{\Delta t/\mu}{1+\frac{\sigma^*\Delta t}{2\mu}}\frac{1}{\Delta y}\left\{E_x^{n}\left(i+\frac{1}{2},j+1,k\right) - E_x^{n}\left(i+\frac{1}{2},j,k\right)\right\}. \qquad (4.48)$$

Note here that, since the isotropic medium is assumed in these equations, the integral numbers of the lattice point (i,j,k) in the electromagnetic constants (σ, ε) are now omitted.

4.1.6 Absorbing Boundary Conditions

Since the analysis of FDTD is fundamentally based on the analysis of the closed region, the absorption boundary conditions, depending on the nature of the analytical region, should be introduced. Let us explain here the problems, specific to the EM-wave absorber analysis.

The intensity of the reflected wave from the EM-wave absorber is usually very weak. Therefore, when conducting a highly accurate analysis, it is necessary to introduce absorption boundary conditions with high absorption performance. As for the absorption boundary conditions, the basic primary and secondary absorption boundary conditions have been proposed from the initial stage of the FDTD analysis [3]. In contrast, Berenger's perfect matched layer (PML) absorption boundary condition has recently been widely introduced [4]. The latter PML absorption boundary is effective, in particular, in the closed-region analysis in the FDTD method, and the absorption boundary wall can be set up as if there was an EM-wave anechoic chamber.

4.1.7 Analysis Model and Boundary Conditions

In order to model the EM-wave absorber, it is required to use the small grid (or lattice size) particularly near the absorber. However, if the lattice size of the

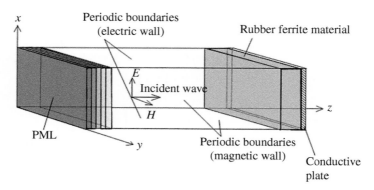

Figure 4.6 Analysis model.

entire analysis region is made small in accordance with the lattice size near the EM-wave absorber, the total number of lattices becomes enormous and it is practically impossible to analyze it. Therefore, it is necessary to introduce the idea of variable lattice size, in which the lattice size in the vicinity of the absorber is made smaller and the lattice size in the free space is made larger.

In the following, as an example of how to establish the FDTD analysis method, let us investigate the problems such as the boundary condition and the convergence of the solutions in the analysis of the EM-wave absorber based on the FDTD method.

As an example of the analysis, a flat rubber ferrite EM-wave absorber is used [5]. This analytical model is shown in Figure 4.6 for the explanation of periodic boundary. In this case, the four sides around the analysis region are surrounded by a pair of electric walls and a pair of magnetic walls, and an analysis is performed in the mirror symmetry structure. That is, as depicted in Figure 4.6, in the rectangular analytical region, the magnetic walls are set on a pair of side walls in the x–z plane parallel to the incident electric field and electrical walls are set on the upper and lower walls of the y–z plane. Figure 4.6 shows the case where the PML is set in the opposite side of the EM-wave absorber [4].

The advantage of this kind of FDTD method is that the analytical region, which has been expanded essentially to the infinite region, can be localized.

The reflection coefficients in the present case can be calculated by introducing the following Gaussian pulse method. First, an incident wave pulse and a reflected wave pulse from the EM-wave absorber to be analyzed are recorded as a time response waveform. Then, after converting to the frequency domain spectrum by the fast Fourier transform (FFT), the reflection coefficient [dB] is given by the following equation:

$$\Gamma = \frac{(E_r)_{FFT}}{(E_i)_{FFT}}$$

$$S(\text{dB}) = 20 \log |\Gamma|.$$

The $(E_r)_{FFT}$ and $(E_i)_{FFT}$ are the Fourier transformed values of the time response waveforms of the reflected wave pulse and the incident wave pulse from the absorber to be analyzed, respectively.

Now, let us here consider the following matters, particularly relevant to the boundary condition and cell size problems:

(a) Behavior of the periodic boundary,
(b) Behavior of the PLM absorption boundary,
(c) Behavior of a variable cell size,
(d) Constitutional dimensions of the absorber and the convergence problem of the solution.

4.1.7.1 Behavior of the Periodic Boundary

First, in order to confirm the validity of this analytical method and the boundary conditions, let us consider a rubber ferrite EM-wave absorber with small square holes. In this analysis, only a quarter of the EM-wave absorber, shown in Figure 4.7, is cut out and used as an analysis model. In Figure 4.7, (a) shows the cut out region of the quarter in (b).

A comparison of the analysis results for the reflection coefficient values in the cases of the ¼ square model in (a) and the whole square model in (b) is shown in Figure 4.7, representing them in circles and crosses, respectively. In this analysis, 300 or more grids are used to satisfy the convergence condition

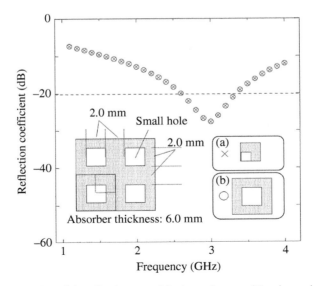

Figure 4.7 Examination of the effectiveness of the boundary condition by analysis. (a) Means the ¼ size square model. (b) Denotes the one square model case.

for analytical values in the previous analysis model with periodic boundaries and the PLM boundary, shown in Figure 4.6.

We can find that the reflection characteristics of both symbols, shown by circles and crosses, perfectly agree with each other, and the introduction of the periodic boundary is extremely rational, even from this example.

4.1.7.2 Behavior of the PLM Absorbing Boundary

Next, let us describe the method to investigate the characteristics of the PML absorption boundary condition. The present analytical model is shown in Figure 4.8. The PML absorption boundary condition is set at one end part in the z direction of the analytical region. On the other side of the PML, a large spatial region is provided. This comes from a countermeasure that the other reflected waves, except for the observed PLM boundary being observed, do not occur. The test conditions for analyzing the present frequency characteristics of the EM-wave absorber with small holes are as follows:

(a) A Gaussian pulse is employed as the incident wave.
(b) The numbers of the PML absorbing boundary wall are composed of 16 layers.
(c) The reflection coefficient at the level of $-120\,dB$ is evaluated here based on the evaluation guideline.

Figure 4.9a shows the reflection coefficients for estimating the performance of the PML absorption boundary wall. In addition, Figure 4.9b shows the time change of the electric field component E_x at the position 10.0 cm away from the PML absorption boundary. In the case of the PML absorption boundary wall composed of 16 layers, it can be estimated that no remarkable reflection is observed, and the absorption boundary wall is sufficiently functional. When the PML absorption boundary wall has eight layers, these characteristics deteriorate, as shown in Figure 4.10a,b, and the reflection coefficient $-120\,dB$ in the evaluation guideline is not satisfied.

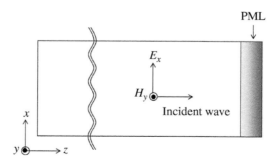

Figure 4.8 Analytical model for investigation of PML absorption boundary condition.

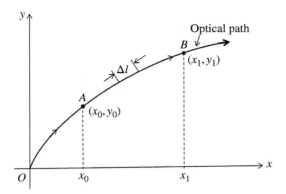

Figure 4.14 Optical path of a medium whose refractive index is continuously changing.

first consider the problem of the light passage time between the points $A(x_0, y_0)$ and $B(x_1, y_1)$. The light is radiated from the origin O, as shown in Figure 4.14.

Let us now assume that the curve of the optical path is given by the expression $y = f(x)$.

In a minute section of the light ray in Figure 4.14, the relation can be expressed as

$$dl = \sqrt{(dx)^2 + (dy)^2} = \sqrt{1 + (y')^2} dx. \tag{4.49}$$

In this case, assuming that the refractive index of the medium is continuously changed and that the light velocity $v(x, y)$ is a function of position on the optical path, the required time from a point A to B is

$$t = \int_{x_0}^{x_1} \frac{\sqrt{1 + (y')^2}}{v(x, y)} dx. \tag{4.50}$$

According to Fermat's principle, light travels along the optical path that minimizes this time.

Hence, if the function $y(x)$ that minimizes time t can be found, the optical path is determined.

In this way, the problem of finding the maximum and minimum values of this integral (referred to as stationary values) is called the variational problem.

Now, if y is a continuous function and satisfies the boundary conditions $y(x_0) = y_0$ and $y(x_1) = y_1$, Eq. (4.50) is generally represented as

$$I(y) = \int_{x_0}^{x_1} F(x, y, y') dx. \tag{4.51}$$

The variable y is a function of x, but it can also be regarded as a variable of $I(y)$. In this sense, y which corresponds to the variable $I(y)$ is called an argument function. Also, $I(y)$ as a function is called "functional."

Let us now assume that $y(x)$ is a stationary function that minimizes the integral $I(y)$. Also, assuming that $\tau(x)$ is an arbitrary function that satisfies the same continuous condition as $y(x)$, and it becomes 0 at the boundaries, we can denote the following function by introducing a parameter α:

$$\bar{y}(x) = y(x) + \alpha\tau(x). \tag{4.52}$$

This expression (4.52) is called a trial function, and also satisfies the condition of becoming 0 on the boundary. Therefore, this means that there are some approximate functions in the vicinity of the stationary function $y(x)$ for various values of α. When considering $\alpha\tau(x)$ as a deviation from a stationary function and it is denoted as $\delta y(= \alpha\tau(x))$, this is called the variation of $y(x)$.

From the explanation, if the functional can be determined in the variational method, it becomes possible to obtain $y(x)$ as a function that renders the stationary value of this integral. By the way, in general, it often becomes difficult to calculate the functional in the variational method mathematically. But, fortunately, the principle of minimum energy exists in the natural world. In the electromagnetic-wave problems, since we are dealing with the problems of regions that are originally based on the potential (energy), we can derive a solution in the form of a variational problem by skillfully using this energy principle.

According to this idea, it becomes possible to solve the variational problem by expressing the functional as a function of energy and utilizing the minimum energy principle, usually established in the energy field.

Taking here an electrostatic field as an example, the energy that the electrostatic field stores in the field can be expressed by the following equation in terms of the relationship $E = -\nabla\phi$:

$$W = I = \frac{1}{2}\epsilon \iiint E^2 dV = \frac{1}{2}\epsilon \iiint (\nabla\phi)^2 dV = \frac{1}{2}\epsilon \iiint \left(\left(\frac{\partial\phi}{\partial x}\right)^2 \right.$$
$$\left. + \left(\frac{\partial\phi}{\partial y}\right)^2 + \left(\frac{\partial\phi}{\partial z}\right)^2 \right) dV, \tag{4.53}$$

where ϕ is the electric potential and ϵ is the dielectric constant.

This expression of the electric field energy is a functional I in the case of an electrostatic field and, consequently, it can be expressed by the functional I in the form of Eq. (4.51).

4.2.1.4 Relationship Between Functional and Laplace Equation

Next, let us investigate the relationship between the functional and the Laplace equation.

Consider the variation δW associated with Eq. (4.53):

$$\delta W = \frac{1}{2}\epsilon\delta \iiint (\nabla\phi)^2 dV. \tag{4.54}$$

By transforming Eq. (4.54) using the relation $\delta\left(\frac{d\phi}{dx}\right) = \frac{d(\delta\phi)}{dx}$ and the relationship of the partial integral $\iiint (\nabla\varphi\nabla\phi + \varphi\Delta\phi)dV = \iint \phi\frac{\partial\varphi}{\partial n}dS$ and then setting this equation to 0, the following equation can be derived:

$$\delta W = -\varepsilon \iiint \delta\phi\Delta\phi \, dV + \varepsilon \iint \delta\phi\frac{\partial\phi}{\partial n} \, dS = 0. \tag{4.55}$$

To satisfy the condition that the given equation holds for any value of $\delta\phi$, the following relationship must be satisfied:

$$\Delta\phi = 0. \tag{4.56}$$

$$\frac{\partial\phi}{\partial n} = 0. \tag{4.57}$$

Expression (4.56) is the Laplace equation, and Expression (4.57) means the natural boundary condition in the boundary region. From this fact it is confirmed that the problem of finding the minimum value of the energy which is stored in the electrostatic field is treated under the existence of Laplace's equation.

4.2.2 Summary of Analytical Procedures

In this section, let us first summarize the analytical procedures in a relatively simple finite element analysis such as a two-dimensional electrostatic field analysis described in the next section.

(a) In the finite element method, as shown in Figure 4.15, the potential surface, which is initially distributed smoothly, is divided into elements, as in Figure 4.5a, and the potential is calculated for each divided element as Figure 4.15b. Needless to say, this method is based on a linear approximation.

(b) Generally, the analysis region is divided into polyhedrons. Usually, a triangle is often used as the shape of the element. Each vertex of a triangle is called a node, and each region of an element is simply called an element.

(c) Then these nodes are numbered consecutively and are expressed as $1e, 2e, 3e, \ldots$, and the coordinates of the nodes are represented as $1e(x_{1e}, y_{1e}), 2e(x_{2e}, y_{2e}), 3e(x_{3e}, y_{3e}), \ldots$.
Also, as for the potential ϕ on the node, the node number is attached as a subscript, such as $\phi_{1e}, \phi_{2e}, \phi_{3e}, \ldots$.
A node whose potential is known is referred to as a known node, and a node whose potential is unknown is referred to as an unknown node.

(d) To represent the potential $\phi^{(e)}$ inside an element with node potentials, the following relations are introduced. One of them is the following linear expression for $\phi^{(e)}$:

$$\phi^{(e)}(x, y) = \alpha_{1e} + \alpha_{2e}x + \alpha_{3e}y,$$

where α_{1e}, α_{2e}, and α_{3e} take different constant values in each element.

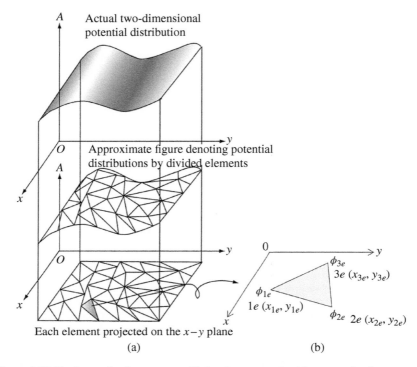

Figure 4.15 The image for the concept of finite element method (an example of two-dimensional analysis).

Another is the following linear expression for $\phi^{(e)}$, for which the interpolation function is denoted as N_{ie}:

$$\phi^{(e)} = N_1\phi_{1e} + N_2\phi_{2e} + N_3\phi_{3e} = \sum_{i=1}^{3} N_{ie}\phi_{ie}.$$

(e) Using these potential expressions, the distribution of electric fields in the region to be analyzed is calculated.

4.2.3 Example of Electrostatic Field Analysis

Let us here explain concretely the finite element analysis in a two-dimensional electrostatic field as an example when a functional is known. The summary of the present analytical procedures of the finite element method is as follows.

(a) First, it is necessary to reformulate a partial differential equation related to a potential into an integral expression, which is called a functional.

(b) As to the integral expression in this functional, it is necessary to impose a condition that the potential is minimized on the basis of the variational principle.

(c) A complementary function is introduced, so that the potential inside the element can be expressed as the value of the potential of the nodes constituting the element.

(d) Consequently, the final simultaneous equations for finding the potential value within an element can be denoted only in terms of the value of node potential without using the potential inside the element.

(e) The potentials in each element of these simultaneous equations can be denoted by a matrix expression, and solutions can be obtained by computer analysis.

Now, assuming that there is no charge in the analysis region, the equation representing the potential is often referred to as the basic equation in the finite element method. In the present case, the basic equation for the electrostatic field of a two-dimensional field is expressed by the following equation:

$$\frac{\partial}{\partial x}\left(\varepsilon\frac{\partial \phi}{\partial x}\right) + \frac{\partial}{\partial y}\left(\varepsilon\frac{\partial \phi}{\partial y}\right) = 0, \tag{4.58}$$

where ϕ is the electric potential and ε is a dielectric constant. This equation is also known as the Laplace equation.

Next, a functional W can be represented as follows, since the basic equation (4.58) corresponds to the Euler equation in the variational method. Once the functional is determined, the equation in the matrix form for all nodes is derived from the condition of the minimum of the functional.

In the present case, the functional, that is, the electric field energy, therefore, is given by the following equation in the same manner as Eq. (4.53) in the previous section:

$$W = I = \frac{1}{2}\int\int_s \left\{\varepsilon\left(\frac{\partial \phi}{\partial x}\right)^2 + \varepsilon\left(\frac{\partial \phi}{\partial y}\right)^2\right\} dx\, dy. \tag{4.59}$$

Since the functional W is a function of the potential ϕ_i at each node i, the condition for minimizing W can be denoted as follows:

$$\frac{dW}{d\phi_i} = 0 \quad (i = 1, 2, \ldots, nt) \ (nt \text{ is the total node numbers}). \tag{4.60}$$

Further, this energy W can be expressed as the sum of the energy $W^{(ne)}$ in each element,

$$\frac{\partial W}{\partial \phi_i} = \frac{\partial W^{(1e)}}{\partial \phi_i} + \cdots - \frac{\partial W^{(ne)}}{\partial \phi_i}. \tag{4.61}$$

Here, ne represents the total number of elements.

As for any element e in the analytical region, the primary triangular elements are adopted, as shown in Figure 4.16. Note here that in the following explanations, only one triangular element is treated, as shown in Figure 4.16b.

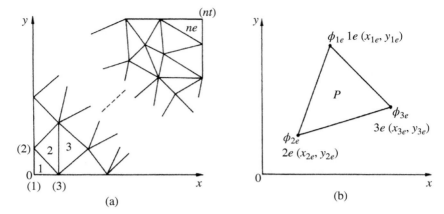

Figure 4.16 Triangular element example.

In the following, we describe a method of the concrete representation of the intra-element potential $\phi^{(e)}$ and the present nodal potential $\phi_{ie}(\phi_{1e}, \phi_{2e}, \phi_{3e})$ with the node coordinates (x, y) according to Eq. (4.59). The purpose of this procedure is to finally obtain simultaneous equations related only to the node coordinates. Using these notations, limited to one element shown in Figure 4.16b, Eq. (4.60) can be rewritten as follows:

$$\frac{\partial W^{(e)}}{\partial \phi_{ie}} = \frac{1}{2} \iint_{s^{(e)}} \left[\varepsilon \frac{\partial}{\partial \phi_{ie}} \left\{ \left(\frac{\partial \phi^{(e)}}{\partial x} \right)^2 + \varepsilon \left(\frac{\partial \phi^{(e)}}{\partial y} \right)^2 \right\} \right] dx\, dy$$

$$= \iint_{s^{(e)}} \varepsilon \left\{ \frac{\partial \phi^{(e)}}{\partial x} \frac{\partial}{\partial \phi_{ie}} \left(\frac{\partial \phi^{(e)}}{\partial x} \right) + \frac{\partial \phi^{(e)}}{\partial y} \frac{\partial}{\partial \phi_{ie}} \left(\frac{\partial \phi^{(e)}}{\partial y} \right)^2 \right\} dx\, dy \quad (4.62)$$

$(= \iint_{s^{(e)}} dx\, dy$ represents the integral in the region $S(e)$ of the element e.)

Here, $W^{(e)}$ means a functional representation that focuses on one element, $\phi^{(e)}$ means the potential distribution within the element, and it is still an unknown function at this point.

Let us here investigate how to rewrite the notation of $\phi^{(e)}$ with the node coordinate expression.

To proceed further with this calculation, we have to assume that the potential $\phi^{(e)}$ within the primary triangular element can be given by the following linear approximation equation, which is expressed by the coordinates of an arbitrary point $P(x, y)$ inside the element:

$$\phi^{(e)}(x, y) = \alpha_{1e} + \alpha_{2e}x + \alpha_{3e}y = [1 \ x \ y] \begin{bmatrix} \alpha_{1e} \\ \alpha_{2e} \\ \alpha_{3e} \end{bmatrix}. \quad (4.63)$$

Here, α_{1e}, α_{2e}, and α_{3e} are constants that are different for each element.

The following is the simultaneous equation yielded by substituting the potentials ϕ_{ie} and coordinates of each node in Figure 4.16b into Eq. (4.63).
Note that the coordinate potentials ϕ_{ie} are included in one element

$$
\begin{bmatrix} \phi_{1e} \\ \phi_{2e} \\ \phi_{3e} \end{bmatrix} = \begin{bmatrix} 1 & x_{1e} & y_{1e} \\ 1 & x_{2e} & y_{2e} \\ 1 & x_{3e} & y_{3e} \end{bmatrix} \begin{bmatrix} \alpha_{1e} \\ \alpha_{2e} \\ \alpha_{3e} \end{bmatrix}. \tag{4.64}
$$

Further, by solving for $\alpha_{1e}, \alpha_{2e}, \alpha_{3e}$ in Eq. (4.64), the following expressions can be obtained:

$$
\alpha_{1e} = \frac{1}{2\Delta^{(e)}} \{ (x_{2e}y_{3e} - x_{3e}y_{2e})\phi_{1e} + (x_{1e}y_{3e} - x_{3e}y_{1e})\phi_{2e} + (x_{1e}y_{2e} - x_{2e}y_{1e})\phi_{3e} \}
$$
$$
\tag{4.65}
$$

$$
\alpha_{2e} = \frac{1}{2\Delta^{(e)}} \{ (y_{2e} - y_{3e})\phi_{1e} + (y_{3e} - y_{1e})\phi_{2e} + (y_{1e} - y_{2e})\phi_{3e} \} \tag{4.66}
$$

$$
\alpha_{3e} = \frac{1}{2\Delta^{(e)}} \{ (x_{3e} - x_{2e})\phi_{1e} + (x_{1e} - x_{3e})\phi_{2e} + (x_{2e} - x_{1e})\phi_{3e} \}, \tag{4.67}
$$

where

$$
2\Delta^{(e)} = \{ (x_{2e}y_{3e} - x_{3e}y_{2e}) + (x_{1e}y_{3e} - x_{3e}y_{1e}) + (x_{1e}y_{2e} - x_{2e}y_{1e}) \}.
$$

By the way, in Eq. (4.62), the functional $W^{(e)}$ is partially differentiated with respect to the potential ϕ_{ie} at each node i. Therefore, if the potential $\phi^{(e)}$ in a given element can be represented by a function of the node potentials ϕ_{ie}, the present analysis can be simplified and made convenient.

Hence, the interpolation function is introduced as this means.

This interpolation function is expressed in terms of $N_{ie}(i_e = 1e, 2e, 3e)$, and $\phi^{(e)}$ is defined as follows:

$$
\phi^{(e)} = N_1\phi_{1e} + N_2\phi_{2e} + N_3\phi_{3e} = \sum_{i=1}^{3} N_{ie}\phi_{ie}. \tag{4.68}
$$

Here, after substituting Eqs. (4.65)–(4.67) into Eq. (4.63) and comparing with Eq. (4.68), the interpolation function N_{ie} can be represented as the following equation:

$$
N_{ie} = \frac{1}{2\Delta^{(e)}} (a_{ie} + b_{ie}x + c_{ie}y), \tag{4.69}
$$

where

$$
a_{ie} = x_{je}y_{ke} - x_{ke}y_{je} \tag{4.70}
$$

$$
b_{ie} = y_{je} - y_{ke} \tag{4.71}
$$

$$
c_{ie} = x_{ke} - x_{je}. \tag{4.72}
$$

Here, *ie*, *je*, *ke* are the node numbers circulating as 1–3 counterclockwise in the element *e* in Figure 4.16b. Now, substituting Eq. (4.68) into Eq. (4.62)

$$\frac{\partial W^{(e)}}{\partial \phi_{ie}} = \iint_{S^{(e)}} \varepsilon \sum_{i=1}^{3} \left(\frac{\partial N_{ie}}{\partial x} \frac{\partial N_{je}}{\partial x} + \frac{\partial N_{ie}}{\partial y} \frac{\partial N_{je}}{\partial y} \right) \phi_{je} dx dy. \tag{4.73}$$

Moreover, from Eq. (4.69), the following relations are obtained:

$$\frac{\partial N_{ie}}{\partial x} = \frac{b_{ie}}{2\Delta^{(e)}} \tag{4.74}$$

$$\frac{\partial N_{ie}}{\partial y} = \frac{c_{ie}}{2\Delta^{(e)}}. \tag{4.75}$$

Substituting Eqs. (4.74) and (4.75) into Eq. (4.73),

$$\frac{\partial W^{(e)}}{\partial \phi_{ie}} = \iint_{S^{(e)}} \varepsilon \sum_{i=1}^{3} \left(\frac{b_{ie}}{2\Delta^{(e)}} \frac{b_{je}}{2\Delta^{(e)}} + \frac{c_{ie}}{2\Delta^{(e)}} \frac{c_{je}}{2\Delta^{(e)}} \right) \phi_{je} \, dx \, dy. \tag{4.76}$$

Also, using the following relation:

$$\iint_{S^{(e)}} dx \, dy = \Delta^{(e)}, \tag{4.77}$$

Eq. (4.76) can be expressed by the following expression:

$$\frac{\partial W^{(e)}}{\partial \phi_{ie}} = \varepsilon \sum_{i=1}^{3} \frac{1}{4\Delta^{(e)}} (b_{ie} \cdot b_{je} + c_{ie} \cdot c_{je}) \phi_{je}. \tag{4.78}$$

Now, after calculating Eq. (4.78) on the three nodes 1*e*, 2*e*, 3*e* in the primary triangular element, by representing Eq. (4.78) in the matrix form,

$$\left\{ \begin{array}{c} \frac{\partial W^{(e)}}{\partial \phi_{1e}} \\ \frac{\partial W^{(e)}}{\partial \phi_{2e}} \\ \frac{\partial W^{(e)}}{\partial \phi_{3e}} \end{array} \right\} = \frac{\varepsilon}{4\Delta^{(e)}} \left[\begin{array}{ccc} b_{1e}b_{1e} + c_{1e}c_{1e} & b_{1e}b_{2e} + c_{1e}c_{2e} & b_{1e}b_{3e} + c_{1e}c_{3e} \\ b_{2e}b_{1e} + c_{2e}c_{1e} & b_{2e}b_{2e} + c_{2e}c_{2e} & b_{2e}b_{3e} + c_{2e}c_{3e} \\ b_{3e}b_{1e} + c_{3e}c_{1e} & b_{3e}b_{2e} + c_{3e}c_{2e} & b_{3e}b_{3e} + c_{3e}c_{3e} \end{array} \right] \left\{ \begin{array}{c} \phi_{1e} \\ \phi_{2e} \\ \phi_{3e} \end{array} \right\}. \tag{4.79}$$

Furthermore, after reduction of Eq. (4.79),

$$\left\{ \begin{array}{c} \frac{\partial W^{(e)}}{\partial \phi_{1e}} \\ \frac{\partial W^{(e)}}{\partial \phi_{2e}} \\ \frac{\partial W^{(e)}}{\partial \phi_{3e}} \end{array} \right\} = \left[\begin{array}{ccc} S_{11}^{(e)} & S_{12}^{(e)} & S_{13}^{(e)} \\ S_{21}^{(e)} & S_{22}^{(e)} & S_{23}^{(e)} \\ S_{31}^{(e)} & S_{32}^{(e)} & S_{33}^{(e)} \end{array} \right] \left\{ \begin{array}{c} \phi_{1e} \\ \phi_{2e} \\ \phi_{3e} \end{array} \right\}. \tag{4.80}$$

In addition, by simplifying further,

$$\left\{ \frac{\partial W^{(e)}}{\partial \phi_{ie}} \right\} = [S_{ij}^{(e)}]\{\phi_{ie}\}. \tag{4.81}$$

Here,

$$S_{ij}^{(e)} = \frac{\varepsilon}{4\Delta^{(e)}}(b_{ie}b_{je} + c_{ie}c_{je}) \quad (i,j = "1,2,3" \text{ between 1 and 3}). \tag{4.82}$$

After all, from Eq. (4.60):

$$[S_{ij}^{(e)}]\{\phi_{ie}\} = \{0\}, \tag{4.83}$$

where

$$\phi = \phi_R - j\phi_i$$
$$S = S_R + jS_i.$$

In this case, when treating a medium with dielectric losses, the potential is naturally represented as a complex number. Since Eq. (4.83) relates only to one triangular element, to calculate all the potentials of the analytical region, it is necessary to solve Expression (4.83), which is taken into account in all elements.

By the way, when the x and y components of the electric field E are expressed by E_x and E_y, respectively, these electric fields have the following relationship with the potential:

$$\left. \begin{array}{l} E_x = -\frac{\partial \phi}{\partial x} \\ E_y = -\frac{\partial \phi}{\partial y} \end{array} \right\}, \tag{4.84}$$

where ϕ is the electric potential.

By first substituting Eq. (4.63) into Eq. (4.84), and then substituting the values of $\alpha_{1e}, \alpha_{2e}, \alpha_{3e}$ from Eqs. (4.65)–(4.67) into Eq. (4.84), the $E_x^{(e)}$ and $E_y^{(e)}$ in the element e are given by the following equation represented by each node potential ϕ_{ie}:

$$\begin{aligned} E_x^{(e)} &= -\frac{1}{2\Delta^{(e)}}(b_{1e}\phi_{1e} + b_{2e}\phi_{2e} + b_{3e}\phi_{3e}) \\ &= -\frac{1}{2\Delta^{(e)}}\{(y_{2e} - y_{3e})\phi_{1e} + (y_{3e} - y_{1e})\phi_{2e} + (y_{1e} - y_{2e})\phi_{3e}\} \end{aligned} \tag{4.85}$$

$$\begin{aligned} E_y^{(e)} &= -\frac{1}{2\Delta^{(e)}}(c_{1e}\phi_{1e} + c_{2e}\phi_{2e} + c_{3e}\phi_{3e}) \\ &= -\frac{1}{2\Delta^{(e)}}\{(x_{3e} - x_{2e})\phi_{1e} + (x_{1e} - x_{3e})\phi_{2e} + (x_{2e} - x_{1e})\phi_{3e}\}. \end{aligned} \tag{4.86}$$

Representing Eqs. (4.85) and (4.86) in terms of a complex notation for ϕ and S, they, of course, become complex equations. Therefore, the magnitude of the electric field $E^{(e)}$ in the element e is determined by synthesizing these expressions. In the abovementioned example, a two-dimensional electrostatic field analysis is considered for introducing the idea of the finite element method.

4.2.4 Application of Electrostatic Field Analysis

As special application examples of FEM, let us briefly discuss the cases in which approximations of the electrostatic field are allowed. For example, in an electric field coupling type wireless power supply device, and an RF capacitive heating type applicator used for cancer thermal therapy, the electrode spacing is extremely narrow compared to the wavelength. In this case, the electrostatic field approximation method was introduced to analyze the electrostatic field distribution [10].

For instance, in an analysis of controlling the electric field distribution, when a cancer treatment applicator by electric field coupling type of RF band (13.45 MHz) is placed in a wide shield room, the advantage of two-dimensional FEM analysis has been confirmed [10]. This can be regarded as a kind of EM-wave absorber in the low-frequency range, which utilizes the property that the electrostatic field converges to the object of a large dielectric constant. That is, given the assumption that the wavelength of 10 MHz frequency is 30 m, and the size of the treatment room is within this wavelength, the behavior of the electromagnetic wave can be treated as a quasi-electrostatic field. Of course, although three-dimensional analyses are most suitable, this type of analysis requires large memory sizes. Based on the present FEM analysis data, a shield room for experimenting with thermotherapy equipment was actually built [11].

4.3 Three-Dimensional Electric Current Potential Method

4.3.1 Outline of the Electric Current Vector Potential Method

In the previous section, the finite element analysis method in the case of two-dimensional electrostatic field analysis was described as an example of using the functional, which is a fundamental variational method. As an example of the electromagnetic-wave analysis introducing the finite element method, the three-dimensional current vector potential method is considered here. In the current potential method, since the analysis is performed using the current potential T and the magnetic scalar potential Ω, it is also called the T–Ω method.

The magnetic field analysis method in the finite element method is largely classified into a magnetic potential method, which is useful when drawing attention to the magnetic field distribution, and a current potential method, which is useful in the case of eddy current analysis.

Here, the method of the current potential is investigated. The reason for considering this method is that it is suitable for analyzing the distribution of the eddy current that occurs when an alternating magnetic field is irradiated to a conductive object, or the like.

At the end of this section, both the eddy current distribution at the time of inductive heating and the eddy current absorbers aimed at controlling unnecessary eddy currents as countermeasures for the re-radiation fields due to eddy currents are discussed. For this reason, first, the analysis method of the three-dimensional current vector potential method is described in detail. In particular, this book discusses in detail the method of discretization, which is said to be difficult to understand. From the viewpoint of the variational method, the method considered in this analysis belongs to direct methods represented by the Galerkin method. The Galerkin method was proposed by Galerkin Boris Grigorievich. The Galerkin method, in this case, is classified as one of the weighted residual methods, and is often used for direct discretization of differential equations.

4.3.2 Basic Equation and Auxiliary Equation

In this section, the basic equations and auxiliary equations necessary for the present analysis is derived. We are aiming at discretizing the electric current vector potential T and the magnetic scalar potential Ω finally, but these analytical processes become somewhat complicated. So, the main theoretical analysis processes are summarized in Table 4.1 with a flowchart.

Let us now express Maxwell's electromagnetic equations in the following form:

$$\nabla \times E = -\frac{\partial B}{\partial t} = -\hat{\mu}\frac{\partial H}{\partial t} \tag{4.87}$$

$$\nabla \times H = J + \frac{\partial D}{\partial t} = \hat{\sigma}E + \hat{\varepsilon}\frac{\partial E}{\partial t} \tag{4.88}$$

$$\nabla \cdot B = 0 \tag{4.89}$$

$$\nabla \cdot D = \rho, \tag{4.90}$$

where E is the field strength, H is magnetic field strength, D is electric flux density, B is magnetic flux density, J is current density, and ρ is the charge density.

In addition, each constitutive relationship equation is given by the following equations:

$$B = \hat{\mu}H \tag{4.91}$$

$$D = \hat{\varepsilon}E \tag{4.92}$$

$$J = \hat{\sigma}E. \tag{4.93}$$

Here, $\hat{\mu}$ is magnetic permeability, $\hat{\varepsilon}$ is permittivity, and $\hat{\sigma}$ is conductivity.

Table 4.1 Flowchart of deriving the electric current vector potential and the magnetic scalar potential.

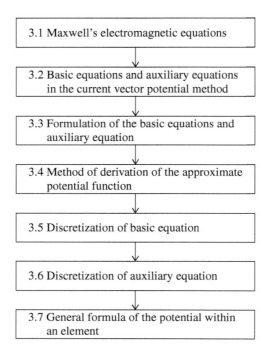

| 3.1 Maxwell's electromagnetic equations |
| 3.2 Basic equations and auxiliary equations in the current vector potential method |
| 3.3 Formulation of the basic equations and auxiliary equation |
| 3.4 Method of derivation of the approximate potential function |
| 3.5 Discretization of basic equation |
| 3.6 Discretization of auxiliary equation |
| 3.7 General formula of the potential within an element |

The basic equation and the auxiliary equation necessary for the present analysis can be derived with the help of these equations. The total magnetic field H of the analyzed area is represented as the sum of the external magnetic field H_0 and eddy current H_e,

$$H = H_0 + H_e. \tag{4.94}$$

Ignoring the displacement current term in Maxwell's equation (4.88), the relationship of the magnetic field H_e generated by the eddy current J_e can be expressed by the following equation:

$$\nabla \times H_e = J_e. \tag{4.95}$$

By applying the vector relation $\nabla \cdot (\nabla \times V) = 0$ in Eq. (4.95), the following equation is obtained:

$$\nabla \cdot J_e = 0. \tag{4.96}$$

Since this relationship of eddy current vectors has the same relation as defining the magnetic vector potential, the present method that uses this equation is called current potential method.

Therefore, it becomes possible to define a current vector potential (electric vector potential) T from Eq. (4.96), and it is expressed by the following equation:

$$J_e = \nabla \times T. \tag{4.97}$$

In addition, by substituting Eq. (4.97) into Eq. (4.95),

$$\nabla \times (H_e - T) = 0. \tag{4.98}$$

Equation (4.98) means that $H_e - T$ is in the conservative system.

Therefore, being the same as the definition of the electric potential, the magnetic potential Ω can be defined as follows:

$$H_e - T = -\nabla\Omega. \tag{4.99}$$

From Eqs. (4.99) and (4.94),

$$H = H_0 + T - \nabla\Omega \tag{4.100}$$

From Eqs. (4.93) and (4.97),

$$E_e = \frac{1}{\hat{\sigma}}\nabla \times T. \tag{4.101}$$

Furthermore, the following relation can be easily obtained:

$$B = \hat{\mu}H = \hat{\mu}(H_0 + T - \nabla\Omega). \tag{4.102}$$

Therefore, by substituting Eqs. (4.101) and (4.102) into Maxwell's equation (4.87), the following expression can be derived:

$$\nabla \times \left(\frac{1}{\hat{\sigma}}(\nabla \times T) \right) = -\frac{\partial}{\partial t}\{\hat{\mu}(H_0 + T - \nabla\Omega)\}. \tag{4.103}$$

This is the "basic equation" of the current vector potential T, which varies with time. By the way, as is clear from expression (4.103), the total number of unknown variables is four, consisting of three components of T and one scalar magnetic potential Ω in this case. Accordingly, from Eq. (4.103), we can obtain only the equation for three components of T. Therefore, another auxiliary equation is necessary.

This auxiliary equation can be derived in terms of the relation $\nabla \cdot B = 0$, as follows:

$$\nabla \cdot \{\hat{\mu}(H_0 + T - \nabla\Omega)\} = 0, \tag{4.104}$$

where $H_0 = H_{0x}i + H_{0y}j + H_{0z}k$, $T = T_e = T_{ex}i + T_{ey}j + T_{ez}k$.

Consequently, solving Eqs. (4.103) and (4.104), we can obtain the current vector potential T and the magnetic scalar potential Ω of the unknown variables, and the eddy current analysis described in this section becomes possible.

4.3.3 Formulations of the Basic and Auxiliary Equations

Here, the formulations mean to express the basic equation (4.103) and the auxiliary equation (4.104) in an integral form, while taking conditions of the analyses into consideration. By obtaining this integral-type expression, it becomes possible to derive the discrete expression described hereafter.

Now, in general, the material constant equations (4.91) and (4.92) are represented by the tensor. But, in the present analysis, since assuming the isotropic medium, the magnetic permeability and conductivity take constant values within each of the present analytical elements.

That is,

$$\hat{\mu} = \begin{bmatrix} \mu_{xx} & \mu_{xy} & \mu_{xz} \\ \mu_{yx} & \mu_{yy} & \mu_{yz} \\ \mu_{zx} & \mu_{zy} & \mu_{zz} \end{bmatrix} = \begin{bmatrix} \mu & 0 & 0 \\ 0 & \mu & 0 \\ 0 & 0 & \mu \end{bmatrix} \tag{4.105}$$

$$\hat{\sigma} = \begin{bmatrix} \sigma_{xx} & \sigma_{xy} & \sigma_{xz} \\ \sigma_{yx} & \sigma_{yy} & \sigma_{yz} \\ \sigma_{zx} & \sigma_{zy} & \sigma_{zz} \end{bmatrix} = \begin{bmatrix} \sigma & 0 & 0 \\ 0 & \sigma & 0 \\ 0 & 0 & \sigma \end{bmatrix}, \tag{4.106}$$

where μ and σ express the values which are limited to one element.

In this potential analysis, a method is adopted in which the potential is discretized to obtain a potential solution. As a means for this, the Galerkin method among the variational methods is applied to the basic equation (4.103) and the auxiliary equation (4.104). As described here, first of all, in the Galerkin method, the integral is expressed using an approximate function T composed of the interpolation function N_i and the potentials (nodal potentials in each node element in this case). And then, it is based on the idea that the residual of this integral is set to zero. Therefore, the Galerkin representations of the basic equation and the auxiliary equation are given by the following integral forms:

$$G_i = \iiint_V N_i \left\{ \nabla \times \left(\frac{1}{\sigma} (\nabla \times T) \right) \right\} dv + \iiint_V N_i \frac{\partial}{\partial t} \{ \mu (H_0 + T - \nabla \Omega) \} \, dv = 0 \tag{4.107}$$

$$G_{di} = \iiint_V N_i \nabla \cdot \{ \mu (H_0 + T - \nabla \Omega) \} \, dv = 0, \tag{4.108}$$

where $i = 1 \sim mt$ (mt is the total number of elements).

Here, the basic equation (4.107) and the auxiliary equation (4.108) should be tried to be further deformed to facilitate discretization. Let us first consider the first term of the basic equation (4.107). On the right-hand side of Eq. (4.107), by applying the partial integration method and the Stokes' theorem to the first term,

$$G_i^1 = \iiint_V N_i \left\{ \nabla \times \left(\frac{1}{\sigma} (\nabla \times T) \right) \right\} dv$$

$$= \iint_S N_i \times n \left(\frac{1}{\sigma} (\nabla \times T) \right) ds - \iiint_V \nabla N_i \times \frac{1}{\sigma} (\nabla \times T) \, dv. \quad (4.109)$$

Here, n is the unit normal vector on the boundary surface S.

In Eq. (4.109), the area integral becomes 0 and disappears. This is due to the following reason. In subsequent analysis procedures, Eq. (4.109) is first discretized, and, finally, we are aiming to obtain a matrix equation. In this analysis, this discretization means the analysis region is divided using the primary tetrahedron element, as described in Section 3.4. This primary representation of the tetrahedral element is based on the volume coordinate concept, and the interpolation coefficient N_i in Eq. (4.109) is also defined in relation to the volume coordinate variables. Therefore, regarding the area integral of the first term on the right-hand side of Eq. (4.109), when considering some two-dimensional plane S, $N_i = 0$.

Also, from another viewpoint it can be found that the area integral term of Eq. (4.109) vanishes. Let us focus on the $(\nabla \times T)$ term of the area component of Eq. (4.109). If the S-plane is assumed to be an xy plane, the z-component J_{ez} of the eddy current perpendicular to this plane cannot exist. When $J_{ez} = 0$, from Eq. (4.97), the x and y components of T are 0, and T_z does not originally exist, so $(\nabla \times T)$ becomes 0. In this way, it is concluded that the area term of Eq. (4.109) becomes 0 in any case. Hence, Eq. (4.107) can be finally represented as follows:

$$G_i = \iiint_V -\frac{1}{\sigma} \nabla N_i \times (\nabla \times T) dv + \iiint_V N_i \frac{\partial}{\partial t} \{ \mu (H_0 + T - \nabla \Omega) \} \, dv = 0.$$

$$(4.110)$$

Further, let us consider the auxiliary equation (4.108). Applying the Gauss divergence theorem and partial integration method, this equation takes the following form:

$$G_{di} = \iiint_V N_i \nabla \cdot \{ \mu (H_0 + T - \nabla \Omega) \} \, dv$$

$$= \iint_S N_i n \cdot \{ \mu (H_0 + T - \nabla \Omega) \} \, ds - \iiint_V \nabla N_i \cdot \{ \mu (H_0 + T - \nabla \Omega) \} \, dv = 0.$$

$$(4.111)$$

For the auxiliary Eq. (4.111), the area integral term becomes 0 when considered similar to that in Eq. (4.109). Therefore, retaining only the second term of the right-hand side of Eq. (4.111), this auxiliary equation is finally represented by the following equation:

$$G_{di} = -\iiint_V \nabla N_i \cdot \{\mu(H_0 + T - \nabla\Omega)\}\, dv = 0. \tag{4.112}$$

4.3.4 Derivation of the Approximate Potential Function

In this section, let us explain specifically the approximate equations for the potential necessary for the discretization of the basic equation (4.110) and the auxiliary equation (4.112).

To discretize these equations, we adopt the primary tetrahedron element as the element shape for dividing the analytical region ("Primary" means that the in-element potential is approximated by a linear equation. Also, the "tetrahedron" suggests that the element consist of four nodes.). The primary tetrahedral element is shown in Figure 4.17a. By the way, when considering an arbitrary point $P(x, y, z)$ in the primary tetrahedral element, it can be noticed that the original primary tetrahedral element can be further decomposed into four primary tetrahedral elements, as shown in Figure 4.17b. In the following analysis, it should be noticed that we focus only on one primary tetrahedral element which consists of four primary tetrahedral elements, shown in Figure 4.17a.

In the configuration of the primary tetrahedral element, when introducing the volume coordinate concept in each divided primary tetrahedron, one can find that the potential of an arbitrary point P in the primary tetrahedron element is denoted by the four node potentials $(T_{1e}, T_{2e}, T_{3e}, T_{4e})$. As shown in Figure 4.17a, a subscript (e) indicates the symbol related to a primary tetrahedron element.

Next, in order to discretize the integrals in Eqs. (4.110) and (4.112) based on the Galerkin method, we need to introduce an approximate potential function, including the interpolation function N_i [12].

In the following, attention is paid to one element (e) in the attempt at discretization. If the coordinates of an arbitrary point P in the element shown in Figure 4.17a is represented by x, y, z, the component of the potential $T^{(e)}$ in the element is given by the following equation.

That is, expressing the approximate potential function $T^{(e)}$ associated with one primary tetrahedron element, using the interpolation function N_{ke}, this scalar potential is usually specified as follows (see Appendix 4.A.1):

$$T^{(e)} = \sum_{k=1}^{4} N_{ke} T_{ke} \quad N_{ke} = \frac{1}{6V^{(e)}}(b_{ke} + c_{ke}x + d_{ke}y + e_{ke}z), \tag{4.113}$$

where N_{ke} is the interpolation function (shape function).

Subscript $ke(k = 1-4)$: Subscript k is an integer related to volume coordinates as shown in Figure 4.17a.

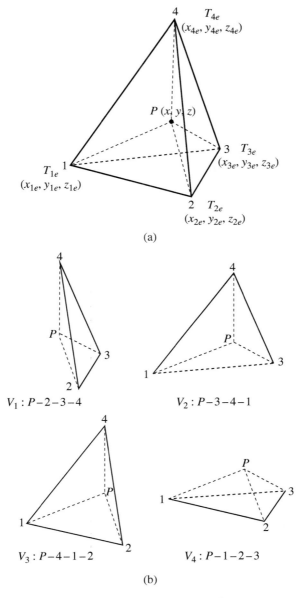

(a)

$V_1 : P-2-3-4$

$V_2 : P-3-4-1$

$V_3 : P-4-1-2$

$V_4 : P-1-2-3$

(b)

Figure 4.17 (a) Primary tetrahedron element. (b) Elements of the primary tetrahedron which are divided into four by point *P*.

$V^{(e)}$: Volume of element (e).

$b_{ke}, c_{ke}, d_{ke}, e_{ke}(k = 1 - 4)$: Constant values are determined by the node coordinates making up element (e).

Hereafter, let us briefly describe the derivation procedure for this approximate function (4.113).

(a) The derivation of this approximate potential function $T^{(e)}$ is based on the concept of volume coordinates. The volume coordinate is a coordinate system consisting of a volume coordinate variable ξ_k, expressed by the following equation, when four volumes in Figure 4.17b are represented by V_1, V_2, V_3, V_4, respectively.
This ξ_k is defined as the following expression:

$$\xi_k = \frac{V_k}{V^{(e)}} = \frac{1}{6V^{(e)}}(b_{ke} + c_{ke}x + d_{ke}y + e_{ke}z). \tag{4.114}$$

Here, $V^{(e)}$ represents the total volume of the primary tetrahedron element (e), and $V_k(k = 1, 2, 3, 4)$ represents each volume of the element (e) divided into four.

(b) Here, the approximate potential function $T^{(e)}$ of an arbitrary point $P(x, y, z)$ of the primary tetrahedral element (e) can be expressed by the following linear combination of ξ_k:

$$T^{(e)} = \alpha_1\xi_1 + \alpha_2\xi_2 + \alpha_3\xi_3 + \alpha_4\xi_4. \tag{4.115}$$

The unknown coefficients $\alpha_1, \alpha_2, \alpha_3, \alpha_4$ in this case can be easily obtained from the following equation. When the point P moves to the node 1 of the primary tetrahedral element (e), as it is clear from the volume relationship indicated in Figure 4.17a and Eq. (4.114), the volumes other than V_1 disappear and, therefore, $(\xi_1, \xi_2, \xi_3, \xi_4)$ can be represented a $\xi_1 = 1, \xi_2 = \xi_3, = \xi_4 = 0$. This means that the potential T_{1e} of node 1 is consistent with α_1. Likewise, the relations between other nodes and the potentials are also given. Accordingly, α_k can be represented by each node potential as follows:

$$T_{1e} = \alpha_1, T_{2e} = \alpha_2, T_{3e} = \alpha_3, T_{4e} = \alpha_4.$$

Here, the suffix ke of T_{ke} means an integer number at each node of the primary tetrahedron element (e). When Eq. (4.115) is represented in a matrix form,

$$T^{(e)} = \begin{bmatrix} \xi_1 & \xi_2 & \xi_3 & \xi_4 \end{bmatrix} \begin{bmatrix} T_1 \\ T_2 \\ T_3 \\ T_4 \end{bmatrix}. \tag{4.116}$$

(c) By substituting Eq. (4.114) into Eq. (4.116), the matrix representation of the potential $T^{(e)}$ at $P(x, y, z)$ within an element is given by the following equation. (Refer to Appendix 4.A.1 for the detailed derivation of the equations.)

$$
T^{(e)} = \frac{1}{6V^{(e)}} \begin{bmatrix} 1 & x & y & z \end{bmatrix} \begin{bmatrix} b_{1e} & b_{2e} & b_{3e} & b_{4e} \\ c_{1e} & c_{2e} & c_{3e} & c_{4e} \\ d_{1e} & d_{2e} & d_{3e} & d_{4e} \\ e_{1e} & e_{2e} & e_{3e} & e_{4e} \end{bmatrix} \begin{bmatrix} T_{1e} \\ T_{2e} \\ T_{3e} \\ T_{4e} \end{bmatrix}
$$

$$
= \begin{bmatrix} 1 & x & y & z \end{bmatrix} \begin{bmatrix} 1 & x_{1e} & y_{1e} & z_{1e} \\ 1 & x_{2e} & y_{2e} & z_{2e} \\ 1 & x_{3e} & y_{3e} & z_{3e} \\ 1 & x_{4e} & y_{4e} & z_{4e} \end{bmatrix}^{-1} \begin{bmatrix} T_{1e} \\ T_{2e} \\ T_{3e} \\ T_{4e} \end{bmatrix} = \left(\sum_{k=1}^{4} N_{ke} T_{ke} \right). \qquad (4.117)
$$

It should be noted here that Eq. (4.113) can be obtained after further calculations of the right-hand side of Eq. (4.117).

Here, the following relationship holds between the constants $b_{ke}, c_{ke}, d_{ke}, e_{ke}$ related to the volume coordinate and the node coordinates x_{ne}, y_{ne}, z_{ne} of the element (e):

$$
\begin{bmatrix} b_{1e} & b_{2e} & b_{3e} & b_{4e} \\ c_{1e} & c_{2e} & c_{3e} & c_{4e} \\ d_{1e} & d_{2e} & d_{3e} & d_{4e} \\ e_{1e} & e_{2e} & e_{3e} & e_{4e} \end{bmatrix} = 6V^{(e)} \begin{bmatrix} 1 & x_{1e} & y_{1e} & z_{1e} \\ 1 & x_{2e} & y_{2e} & z_{2e} \\ 1 & x_{3e} & y_{3e} & z_{3e} \\ 1 & x_{4e} & y_{4e} & z_{4e} \end{bmatrix}^{-1}, \qquad (4.118)
$$

where $V^{(e)}$ is the total volume of the primary tetrahedral element

$$
V^{(e)} = \frac{1}{6} \begin{vmatrix} 1 & x_{1e} & y_{1e} & z_{1e} \\ 1 & x_{2e} & y_{2e} & z_{2e} \\ 1 & x_{3e} & y_{3e} & z_{3e} \\ 1 & x_{4e} & y_{4e} & z_{4e} \end{vmatrix}. \qquad (4.119)
$$

The meaning of Eq. (4.117) is that the potential of a point $P(x, y, z)$ in a primary tetrahedron element can be determined from the node coordinates (x_{ke}, y_{ke}, z_{ke}) and potentials $(T_{1e}, T_{2e}, T_{3e}, T_{4e})$ of each node.

Now, Eq. (4.113) is an approximate expression of the scalar potential. On the other hand, the potential T in Eq. (4.110) is a vector expression. Accordingly, the approximate equation for the vector potential T, corresponding to

Eq. (4.110), is defined as follows:

$$T^{(e)} = iT_x + jT_y + kT_z \quad (i, j, k : \text{Unit vector}). \tag{4.120}$$

With reference to Eq. (4.113), the x, y, z components of the present vector potential are therefore given by the following equations[2]:

$$T_x^{(e)} = \sum_{i=1}^{4} N_{ie} T_{xie} \tag{4.121}$$

$$T_y^{(e)} = \sum_{i=1}^{4} N_{ie} T_{yie} \tag{4.122}$$

$$T_z^{(e)} = \sum_{i=1}^{4} N_{ie} T_{zie}. \tag{4.123}$$

Similarly, the following relation is introduced between the interpolation function N_{ie} and the magnetic scalar potential $\Omega^{(e)}$:

$$\Omega^{(e)} = \sum_{i=1}^{4} N_{ie} \Omega_{ie}. \tag{4.124}$$

Here, $\Omega^{(e)}$ represents the magnetic scalar potential within the element (e).

4.3.5 Discretization of the Basic Equation

In this section, we first treat the discretization of the basic equation (4.110), previously described in Section 4.3.3. Now, let us here rewrite Eq. (4.110) again,

$$G_i = \iiint_V -\frac{1}{\sigma} \nabla N_i \times (\nabla \times T) \, dv$$
$$+ \iiint_V N_i \frac{\partial}{\partial t} \{\mu(H_0 + T - \nabla \Omega)\} \, dv = 0. \quad (4.110 \text{ repeated})$$

In the following, we consider the discretization of each term in the basic equation of (4.110) in order. Since this analysis process is complicated, its flow chart is shown in Table 4.2. Based on the equations derived so far, let us start the discretization in the basic equation (4.110) in the order shown in the following Table 4.2.

4.3.5.1 The First Term on the Right Side of Eq. (4.110)

First, let us consider the discretization of the first term on the right-hand side of the expression (4.110). In this case, the part of the integrand, excluding the

---

2 The subscript k in Eq. (4.113), which is used in connection with the appendix, is now formally rewritten as i in accordance with the interpolation coefficient expressions of Eqs. (4.107) and (4.108). That is, $i = 1$–4 (element numbers).

Table 4.2 Discretization processes of the basic equation.

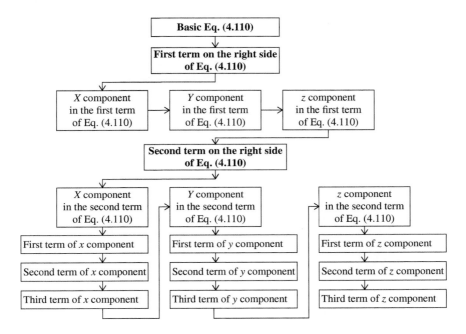

coefficient part, is expressed as follows.

$$
\nabla N_i \times (\nabla \times \mathbf{T}) = \left\{ \frac{\partial N_i}{\partial y} \left(\frac{\partial T_y}{\partial x} - \frac{\partial T_x}{\partial y} \right) - \frac{\partial N_i}{\partial z} \left(\frac{\partial T_x}{\partial z} - \frac{\partial T_z}{\partial x} \right) \right\} \mathbf{i}
$$
$$
+ \left\{ \frac{\partial N_i}{\partial z} \left(\frac{\partial T_z}{\partial y} - \frac{\partial T_y}{\partial z} \right) - \frac{\partial N_i}{\partial x} \left(\frac{\partial T_y}{\partial x} - \frac{\partial T_x}{\partial y} \right) \right\} \mathbf{j}
$$
$$
+ \left\{ \frac{\partial N_i}{\partial x} \left(\frac{\partial T_x}{\partial z} - \frac{\partial T_z}{\partial x} \right) - \frac{\partial N_i}{\partial y} \left(\frac{\partial T_z}{\partial y} - \frac{\partial T_y}{\partial z} \right) \right\} \mathbf{k}.
$$

$$(4.125)$$

Now, the derivatives in the first term of Eq. (4.125) before the parentheses ( ) are computed using $N_{ie}(= N_{ke})$ in Eq. (4.113).[3]

$$
\frac{\partial N_i}{\partial x} = \frac{\partial}{\partial x} \left\{ \frac{1}{6V^{(e)}} (b_{ie} + c_{ie}x + d_{ie}y + e_{ie}z) \right\} = \frac{c_{ie}}{6V^{(e)}} \tag{4.126}
$$

$$
\frac{\partial N_i}{\partial y} = \frac{\partial}{\partial y} \left\{ \frac{1}{6V^{(e)}} (b_{ie} + c_{ie}x + d_{ie}y + e_{ie}z) \right\} = \frac{d_{ie}}{6V^{(e)}} \tag{4.127}
$$

---

3 Here, as apparent from the relationship of Expression (4.118), the following coefficients $b_{ie}, c_{ie}, d_{ie} - - -, d_{je}, e_{je} - - -$, etc., are determined from the node coordinate values.

$$\frac{\partial N_i}{\partial z} = \frac{\partial}{\partial z}\left\{\frac{1}{6V^{(e)}}(b_{ie} + c_{ie}x + d_{ie}y + e_{ie}z)\right\} = \frac{e_{ie}}{6V^{(e)}}. \tag{4.128}$$

Note here that e is often omitted for simplicity in the following analysis except the case where it is necessary.

4.3.5.2 x Component in the First Term of the Basic Equation (4.110)

First, from the relationship of Eq. (4.125), the x component of the first term of Eq. (4.110) can be calculated as follows:

$$\left[-\frac{1}{\sigma}\iiint_V \nabla N_i \times (\nabla \times \boldsymbol{T})dv\right]_x$$

$$= \iiint_V -\frac{1}{\sigma}\left\{\frac{\partial N_i}{\partial y}\left(\frac{\partial T_y}{\partial x} - \frac{\partial T_x}{\partial y}\right) - \frac{\partial N_i}{\partial z}\left(\frac{\partial T_x}{\partial z} - \frac{\partial T_z}{\partial x}\right)\right\}dv$$

$$= \iiint_V -\frac{1}{\sigma}\left\{\frac{d_{ie}}{6V^{(e)}}\left(\frac{\partial \sum_{j=1}^4 N_j T_{yj}}{\partial x} - \frac{\partial \sum_{j=1}^4 N_j T_{xj}}{\partial y}\right)\right.$$

$$\left. - \frac{e_{ie}}{6V^{(e)}}\left(\frac{\partial \sum_{j=1}^4 N_j T_{xj}}{\partial z} - \frac{\partial \sum_{j=1}^4 N_j T_{zj}}{\partial x}\right)\right\}dv$$

$$= \iiint_V -\frac{1}{\sigma}\sum_{j=1}^4\left\{\frac{d_{ie}}{6V^{(e)}}\left(\frac{c_{je}}{6V^{(e)}}T_{yj} - \frac{d_{je}}{6V^{(e)}}T_{xj}\right)\right.$$

$$\left. - \frac{e_{ie}}{6V^{(e)}}\left(\frac{e_{je}}{6V^{(e)}}T_{xj} - \frac{c_{je}}{6V^{(e)}}T_{zj}\right)\right\}dv$$

$$= -\frac{1}{36V^{(e)2}\sigma}\sum_{j=1}^4(-d_{ie}d_{je}T_{xj} + d_{ie}c_{je}T_{yj} - e_{ie}e_{je}T_{xj} + c_{je}e_{ie}T_{zj}) \times \iiint_V dv$$

$$= \frac{1}{36V^{(e)}\sigma}\sum_{j=1}^4\{(d_{ie}d_{je} + e_{ie}e_{je})T_{xj} - d_{ie}c_{je}T_{yj} - e_{ie}c_{je}T_{zj}\}. \tag{4.129}$$

In Eq. (4.129), superscripts and subscripts (e) for derivatives with respect to the x, y, z variables *in* the potential equations (4.121)–(4.123) are omitted.

Also, each term obtained by differentiating N_i with respect to the variables x, y, z is no longer related to the variables x, y, z and becomes associated with each coordinate value of the node. Therefore, the integer value i, which represents the element number, is replaced by an integer value j associated with the node number.

Further, note that the constants $(d_{ie}, d_{je}, c_{je}, \ldots)$ related to N_{ie} and the newly introduced N_{je} that are associated with the node values are explicitly indicated by the addition of the suffix e.

The purpose of rewriting Eq. (4.129) is to simplify its expression and to finally rewrite the matrix, expressed by the potential of each element, with the matrix expression of the nodal potential in order to be able to easily analyze some model of analysis.

4.3.5.3 y Component in the First Term of Basic Equation (4.110)

Similarly, in Eq. (4.110), the y component of the first term on the right-hand side,

$$\left[-\frac{1}{\sigma} \iiint_V \nabla N_i \times (\nabla \times \boldsymbol{T}) dv \right]_y$$

$$= \iiint_V -\frac{1}{\sigma} \left\{ \frac{\partial N_i}{\partial z} \left(\frac{\partial T_z}{\partial y} - \frac{\partial T_y}{\partial z} \right) - \frac{\partial N_i}{\partial x} \left(\frac{\partial T_y}{\partial x} - \frac{\partial T_x}{\partial y} \right) \right\} dv$$

$$= \frac{1}{36 V^{(e)} \sigma} \sum_{j=1}^{4} \{ -c_{ie} d_{je} T_{xj} + (c_{ie} c_{je} + e_{ie} e_{je}) T_{yj} - e_{ie} d_{je} T_{zj} \}. \qquad (4.130)$$

(See Appendix 4.A.2.)

4.3.5.4 z Component in the First Term of the Basic Equation (4.110)

In the same way, in Eq. (4.110), the z component of the first term on the right-hand side is

$$\left[-\frac{1}{\sigma} \iiint_V \nabla N_i \times (\nabla \times \boldsymbol{T}) dv \right]_z$$

$$= \iiint_V -\frac{1}{\sigma} \left\{ \frac{\partial N_i}{\partial x} \left(\frac{\partial T_x}{\partial z} - \frac{\partial T_z}{\partial x} \right) - \frac{\partial N_i}{\partial y} \left(\frac{\partial T_z}{\partial y} - \frac{\partial T_y}{\partial z} \right) \right\} dv$$

$$= \frac{1}{36 V^{(e)} \sigma} \sum_{j=1}^{4} \{ -c_{ie} e_{je} T_{xj} - d_{ie} e_{je} T_{yj} + (c_{ie} c_{je} + d_{ie} d_{je}) T_{zj} \}. \qquad (4.131)$$

(See Appendix 4.A.3.)

4.3.5.5 The Second Term on the Right Side of Eq. (4.110)

Next, let us consider the second term on the right-hand side of the basic equation (4.110), represented by the integral form. Here, the time derivative in the second term on the right-hand side is replaced by $j\omega$. In addition, grad Ω can be expressed as follows:

$$\nabla \Omega = \frac{\partial \Omega}{\partial x} \boldsymbol{i} + \frac{\partial \Omega}{\partial y} \boldsymbol{j} + \frac{\partial \Omega}{\partial z} \boldsymbol{k}. \qquad (4.132)$$

Since the second term of Eq. (4.110) contains an applied magnetic field (external magnetic field) $\boldsymbol{H}_0 = H_{0x} \boldsymbol{i} + H_{0y} \boldsymbol{j} + H_{0z} \boldsymbol{k}$, the x component of the second term in the integral of Eq. (4.110) is expressed by the following equation:

4.3.5.6 x Component of the Second Term on the Right Side of the Basic Equation (4.110)

$$\left[\iiint_V N_i \frac{\partial}{\partial t}\{\mu(H_0 + T - \nabla\Omega)\}\, dv\right]_x$$

$$= j\omega\mu\iiint_V N_i H_{0x}\, dv + j\omega\mu\iiint_V N_i \sum_{j=1}^{4} N_j T_{xj}\, dv - j\omega\mu\iiint_V N_i \frac{\sum_{j=1}^{4} N_j \Omega_j}{\partial x}\, dv$$

$$= j\omega\mu H_{0x}\iiint_V N_i\, dv + j\omega\mu\sum_{j=1}^{4} T_{xj}\iiint_V N_i N_j\, dv - j\omega\mu\sum_{j=1}^{4}\iiint_V N_i \frac{c_{je}}{6V^{(e)}}\Omega_j\, dv. \quad (4.133)$$

$\underbrace{\hspace{9cm}}$

X component of the second term on the right side in the basic equation (4.110)

This expression represents the result of executing the volume integral in Eq. (4.110) for the x component. To calculate each term in Eq. (4.133), we need to introduce here the volume integral formula in the primary tetrahedral element. This volume integral formula is given by

$$\iiint_V N_1^a N_2^b N_3^c N_4^d\, dx\, dy\, dz = \frac{a!b!c!d!}{(a+b+c+d+3)!}\times 6V^{(e)}. \quad (4.134)$$

4.3.5.7 The First Term of x Component in Eq. (4.133)

When applying the volume integral equation (4.134) to Eq. (4.133), the first term on the right side of Eq. (4.133) is expressed as follows:

$$j\omega\mu H_{0x}\iiint_V N_i\, dv = j\omega\mu H_{0x}\frac{1!0!0!0!}{(1+0+0+0+3)!}\times 6V^{(e)}$$

$$= j\omega\mu H_{0x}\frac{1}{4!}6V^{(e)} = j\omega\mu H_{0x}\frac{V^{(e)}}{4}. \quad (4.135)$$

4.3.5.8 The Second Term of the x Component in Eq. (4.133)

Next, the second term on the right-hand side of Eq. (4.133) is classified on the basis of the following conditions:

In the case of $i = j$,

$$j\omega\mu\sum_{j=1}^{4} T_{xj}\iiint_V N_i^2\, dv = j\omega\mu\sum_{j=1}^{4} T_{xj}\frac{2!0!0!0!}{(2+0+0+0+3)!}\times 6V^{(e)}$$

$$= j\omega\mu\sum_{j=1}^{4} T_{xj}\frac{2!}{5!}6V^{(e)} = \frac{j\omega\mu V^{(e)}}{10}\sum_{j=1}^{4} T_{xj}. \quad (4.136)$$

In the case of $i \neq j$,

$$j\omega\mu\sum_{j=1}^{4} T_{xj}\iiint_V N_i N_j\, dv = j\omega\mu\sum_{j=1}^{4} T_{xj}\frac{1!1!0!0!}{(1+1+0+0+3)!}\times 6V^{(e)}$$

$$= j\omega\mu\sum_{j=1}^{4} T_{xj}\frac{1!}{5!}6V^{(e)} = \frac{j\omega\mu V^{(e)}}{20}\sum_{j=1}^{4} T_{xj}. \quad (4.137)$$

Accordingly, the second term of Eq. (4.133) is generally expressed using the Kronecker delta δ_{ij} as follows:

$$j\omega\mu \sum_{j=1}^{4} T_{xj} \iiint_V N_i N_j \, dv = \frac{j\omega\mu V^{(e)}}{20}(1 + \delta_{ij}) \sum_{j=1}^{4} T_{xj} \begin{cases} i = j & \delta_{ij} = 1 \\ i \neq j & \delta_{ij} = 0 \end{cases}.$$

$$(4.138)$$

4.3.5.9 The Third Term of the *x* Component in Eq. (4.133)

In the same manner, the third term of the x component in Eq. (4.133):

$$j\omega\mu \sum_{j=1}^{4} \iiint_V N_i \frac{c_{je}}{6V^{(e)}} \Omega_{je} dv = -\frac{j\omega\mu}{6V^{(e)}} \sum_{j=1}^{4} c_{je}\Omega_{je} \iiint_V N_i dv$$

$$= -\frac{j\omega\mu}{6V^{(e)}} \sum_{j=1}^{4} c_{je}\Omega_{je} \frac{1!0!0!0!}{(1+0+0+0+3)!} \times 6V^{(e)}$$

$$= -\frac{j\omega\mu}{6V^{(e)}} \sum_{j=1}^{4} c_{je}\Omega_{je} \frac{6V^{(e)}}{4!} = -\frac{j\omega\mu}{24} \sum_{j=1}^{4} c_{je}\Omega_{je}.$$

$$(4.139)$$

These were the expressions for the *x* component of the second term of Eq. (4.110). Similarly, the components *y* and *z* of the second term of Eq. (4.110) can be derived. From these calculated results, if G_i in Eq. (4.110) is represented by the *x, y, z* component, it can be summarized as follows:

$$G_{ix} = \frac{1}{36V^{(e)}\sigma} \sum_{j=1}^{4} \{(d_{ie}d_{je} + e_{ie}e_{je})T_{xj} - d_{ie}c_{je}T_{yj} - e_{ie}c_{je}T_{zj}\} + j\omega\mu H_{0x}\frac{V^{(e)}}{4}$$

$$+ \frac{j\omega\mu V^{(e)}}{20}(1 + \delta_{ij}) \sum_{j=1}^{4} T_{xj} - \frac{j\omega\mu}{24} \sum_{j=0}^{4} c_{je}\Omega_{je}$$

$$= \sum_{j=1}^{4} \left\{ \frac{1}{9V^{(e)}\sigma}(d_{ie}d_{je} + e_{ie}e_{je}) + \frac{j\omega\mu V^{(e)}}{5}(1 + \delta_{ij}) \right\} T_{xj} - \sum_{j=1}^{4} \frac{d_{ie}c_{je}}{9V^{(e)}\sigma} T_{yj}$$

$$- \sum_{j=1}^{4} \frac{e_{ie}c_{je}}{9V^{(e)}\sigma} T_{zj} - \frac{j\omega\mu}{6} \sum_{j=1}^{4} c_{je}\Omega_{je} + j\omega\mu H_{0x} V^{(e)} = 0. \qquad (4.140)$$

$$G_{iy} = \frac{1}{36V^{(e)}\sigma} \sum_{j=1}^{4} \{-c_{ie}d_{je}T_{xj} + (c_{ie}c_{je} + e_{je}e_{ie})T_{yj} - e_{ie}d_{je}T_{zj}\} + j\omega\mu H_{0y}\frac{V^{(e)}}{4}$$

$$+ \frac{j\omega\mu V^{(e)}}{20}(1 + \delta_{ij}) \sum_{j=1}^{4} T_{yj} - \frac{j\omega\mu}{24} \sum_{j=1}^{4} d_{je}\Omega_{je}$$

$$= -\sum_{j=1}^{4} \frac{c_{ie}d_{je}}{9V^{(e)}\sigma} T_{xj} + \sum_{j=1}^{4} \left\{ \frac{1}{9V^{(e)}\sigma}(c_{ie}c_{je} + e_{je}e_{ie}) + \frac{j\omega\mu V^{(e)}}{5}(1 + \delta_{ij}) \right\} T_{yj}$$

$$- \sum_{j=1}^{4} \frac{e_{ie}d_{je}}{9V^{(e)}\sigma} T_{zj} - \frac{j\omega\mu}{6} \sum_{j=1}^{4} d_{je}\Omega_{je} + j\omega\mu H_{0y} V^{(e)} = 0. \qquad (4.141)$$

$$G_{iz} = \frac{1}{36V^{(e)}\sigma} \sum_{j=1}^{4} \{-c_{ie}e_{je}T_{xj} - d_{ie}e_{je}T_{yj} + (c_{je}c_{ie} + d_{ie}d_{je})T_{zj}\} + j\omega\mu H_{0z}\frac{V^{(e)}}{4}$$

$$+ \frac{j\omega\mu V^{(e)}}{20}(1 + \delta_{ij}) \sum_{j=1}^{4} T_{zj} - \frac{j\omega\mu}{24} \sum_{j=1}^{4} e_{je}\Omega_{je}$$

$$= -\sum_{j=1}^{4} \frac{c_{ie}e_{je}}{9V^{(e)}\sigma} T_{xj} - \sum_{j=1}^{4} \frac{d_{ie}e_{je}}{9V^{(e)}\sigma} T_{yj} + \sum_{j=1}^{4} \left\{ \frac{1}{9V^{(e)}\sigma}(c_{je}c_{ie} + d_{ie}d_{je}) \right.$$

$$\left. + \frac{j\omega\mu V^{(e)}}{20}(1 + \delta_{ij}) \right\} T_{zj}$$

$$- \frac{j\omega\mu}{6} \sum_{j=1}^{4} e_{je}\Omega_{je} + j\omega\mu H_{0z} V^{(e)} = 0. \qquad (4.142)$$

4.3.6 Discretization of the Auxiliary Equation

Next, we consider the method of discretization of the auxiliary equation in (4.112). Let us again rewrite the auxiliary equation in (4.112) as

$$G_{di} = -\iiint_{V} \nabla N_i \cdot \{\mu(\boldsymbol{H}_0 + \boldsymbol{T} - \nabla\Omega)\} dv = 0. \quad (4.112 \text{ repeated})$$

Here, $\mu(\boldsymbol{H}_0 + \boldsymbol{T} - \nabla\Omega)$ can be denoted by components as follows:

$$\mu(\boldsymbol{H}_0 + \boldsymbol{T} - \nabla\Omega) = \mu \left\{ \left(H_{0x} + T_x - \frac{\partial\Omega}{\partial x} \right) \boldsymbol{i} + \left(H_{0y} + T_y - \frac{\partial\Omega}{\partial y} \right) \boldsymbol{j} \right\}$$

$$+ \left(H_{0z} + T_z - \frac{\partial\Omega}{\partial z} \right) \boldsymbol{k} \right\}. \qquad (4.143)$$

Accordingly, from Eq. (4.143) and the relations expressed in scalar form in the previous equations (4.126)–(4.128),

$$\nabla N_i \cdot \{\mu(\boldsymbol{H}_0 + \boldsymbol{T} - \nabla\Omega)\}$$

$$= \mu \frac{\partial N_i}{\partial x} \left(H_{0x} + T_x - \frac{\partial\Omega}{\partial x} \right) + \mu \frac{\partial N_i}{\partial y} \left(H_{0y} + T_y - \frac{\partial\Omega}{\partial y} \right)$$

$$+ \mu \frac{\partial N_i}{\partial z} \left(H_{0z} + T_z - \frac{\partial\Omega}{\partial z} \right)$$

$$
= \frac{\mu c_{ie}}{6V^{(e)}} \left(H_{0x} + \sum_{j=1}^{4} N_j T_{xj} - \sum_{j=1}^{4} \frac{c_{je}}{6V^{(e)}} \Omega_j \right)
$$

$$
+ \frac{\mu d_{ie}}{6V^{(e)}} \left(H_{0y} + \sum_{j=1}^{4} N_j T_{yj} - \sum_{j=1}^{4} \frac{d_{je}}{6V^{(e)}} \Omega_j \right)
$$

$$
+ \frac{\mu e_{ie}}{6V^{(e)}} \left(H_{0z} + \sum_{j=1}^{4} N_j T_{zj} - \sum_{j=1}^{4} \frac{e_{je}}{6V^{(e)}} \Omega_j \right). \tag{4.144}
$$

In the following, the calculations are performed for each of the x, y, z components in Eq. (4.144).

4.3.6.1 x Component in Eq. (4.112)

First, let consider the x component in Eq. (4.112).

$$
\left[-\iiint_V \nabla N_i \cdot \{ \mu (\boldsymbol{H}_0 + \boldsymbol{T} - \nabla \Omega) \} \, dv \right]_x
$$

$$
= -\iiint_V \frac{\mu c_{ie}}{6V^{(e)}} \left(H_{0x} + \sum_{j=1}^{4} N_j T_{xj} - \sum_{j=1}^{4} \frac{c_{je}}{6V^{(e)}} \Omega_j \right) dv
$$

$$
= -\frac{\mu c_{ie}}{6V^{(e)}} \left\{ V^{(e)} H_{0x} + \frac{1!0!0!0!}{(1+0+0+0+3)!} 6V^{(e)} \sum_{j=1}^{4} T_{xj} - \sum_{j=1}^{4} \frac{c_{je}}{6} \Omega_j \right\}
$$

$$
= -\frac{\mu c_{ie}}{6} H_{0x} - \frac{\mu c_{ie}}{24} \sum_{j=1}^{4} T_{xj} + \frac{\mu c_{ie}}{36} \sum_{j=1}^{4} c_{je} \Omega_j
$$

$$
= \frac{\mu c_{ie}}{6} \left(-H_{0x} - \frac{1}{4} \sum_{j=1}^{4} T_{xj} + \frac{1}{6V^{(e)}} \sum_{j=1}^{4} c_{je} \Omega_j \right). \tag{4.145}
$$

Similarly, the y and z components of Eq. (4.144) can also be obtained.

Consequently, if we unify the x, y, and z components of Eq. (4.144) into a single expression, the discretization form of the auxiliary equation (4.112) can be given as follows:

$$
G_{di} = \frac{\mu c_{ie}}{6} \left(-H_{0x} - \frac{1}{4} \sum_{j=1}^{4} T_{xj} + \frac{1}{6V^{(e)}} \sum_{j=1}^{4} c_{je} \Omega_j \right)
$$

$$
+ \frac{\mu d_{ie}}{6} \left(-H_{0y} - \frac{1}{4} \sum_{j=1}^{4} T_{yj} + \frac{1}{6V^{(e)}} \sum_{j=1}^{4} d_{je} \Omega_j \right)
$$

$$
+ \frac{\mu e_{ie}}{6} \left(-H_{0z} - \frac{1}{4} \sum_{j=1}^{4} T_{zj} + \frac{1}{6V^{(e)}} \sum_{j=1}^{4} e_{je} \Omega_j \right)
$$

$$= -\frac{\mu}{4}\left(c_{ie}\sum_{j=1}^{4}T_{xj} + d_{ie}\sum_{j=1}^{4}T_{yj} + e_{ie}\sum_{j=1}^{4}T_{zj}\right) - \mu(c_{ie}H_{0x} + d_{ie}H_{0y} + e_{ie}H_{0z})$$

$$+ \frac{\mu}{6V^{(e)}}\sum_{j=1}^{4}(c_{ie}c_{je} + d_{ie}d_{je} + e_{ie}e_{je})\Omega_j = 0. \tag{4.146}$$

From these calculation results, a simultaneous linear equation with the x, y, z component of the vector potential $T^{(e)}$ and the scalar potential Ω as the unknowns can be derived.

4.3.7 General Potential Equation in Elements

In the previous section, focusing on the element shown in Figure 4.17, the x, y, z components of the discretized vector current potential and the magnetic scalar potential could be derived. In this section, let us consider integrating all the calculated components of the basic equation and auxiliary equation into one expression, which is denoted by a matrix form.

This result is represented in a matrix form in the following expression:

$$\begin{bmatrix} G_{ij}^{xx} & G_{ij}^{xy} & G_{ij}^{xz} & G_{ij}^{x\Omega} \\ G_{ji}^{yx} & G_{ij}^{yy} & G_{ij}^{yz} & G_{ij}^{y\Omega} \\ G_{ji}^{zx} & G_{ii}^{zy} & G_{ij}^{zz} & G_{ij}^{z\Omega} \\ G_{ji}^{\Omega x} & G_{ji}^{\Omega y} & G_{ji}^{\Omega z} & G_{ij}^{\Omega\Omega} \end{bmatrix} \begin{bmatrix} T_{xj} \\ T_{yj} \\ T_{zj} \\ \Omega_j \end{bmatrix} = \begin{bmatrix} W_i \\ X_i \\ Y_i \\ Z_i \end{bmatrix}, \tag{4.147}$$

where i (element numbers), $j = 1$–4 (node numbers in an element).

Here, for example, the superscript xy of G means the y component of Eq. (4.148) of the x component in the basic equation (4.110).

Here,

$$G_{ij}^{xx} = \sum_{j=1}^{4}\left\{\frac{1}{9V^{(e)}\sigma}(d_{ie}d_{je} + e_{ie}e_{je}) + \frac{j\omega\mu V^{(e)}}{5}(1 + \delta_{ij})\right\}, \quad G_{ij}^{xy} = -\sum_{j=1}^{4}\frac{d_{ie}c_{je}}{9V^{(e)}\sigma}$$

$$G_{ij}^{xy} = -\sum_{j=1}^{4}\frac{d_{ie}c_{je}}{9V^{(e)}\sigma}, \quad G_{ij}^{xz} = -\sum_{j=1}^{4}\frac{e_{ie}c_{je}}{9V^{(e)}\sigma}$$

$$G_{ij}^{x\Omega} = -\frac{j\omega\mu}{6}\sum_{j=1}^{4}c_{je}, \quad G_{ij}^{yx} = -\sum_{j=1}^{4}\frac{c_{ie}d_{je}}{9V^{(e)}\sigma}$$

$$G_{ij}^{yy} = \sum_{j=1}^{4}\left\{\frac{1}{9V^{(e)}\sigma}(c_{ie}c_{je} + e_{je}e_{ie}) + \frac{j\omega\mu V^{(e)}}{5}(1 + \delta_{ij})\right\},$$

$$G_{ij}^{yz} = -\sum_{j=1}^{4}\frac{e_{ie}d_{je}}{9V^{(e)}\sigma}$$

$$G_{ij}^{y\Omega} = -\frac{j\omega\mu}{6}\sum_{j=1}^{4} d_{je}, \quad G_{ij}^{zx} = -\sum_{j=1}^{4}\frac{c_{ie}e_{je}}{9V^{(e)}\sigma}$$

$$G_{ij}^{zy} = -\sum_{j=1}^{4}\frac{d_{ie}e_{je}}{9V^{(e)}\sigma},$$

$$G_{ij}^{zz} = \sum_{j=1}^{4}\left\{\frac{1}{9V^{(e)}\sigma}(c_{je}c_{ie} + d_{ie}d_{je}) + \frac{j\omega\mu V^{(e)}}{20}(1 + \delta_{ij})\right\}$$

$$G_{ij}^{z\Omega} = -\frac{j\omega\mu}{6}\sum_{j=1}^{4} e_{je}, \quad G_{ij}^{\Omega x} = -\frac{\mu}{4}\sum_{j=1}^{4} c_{ie}$$

$$G_{ij}^{\Omega y} = -\frac{\mu}{4}\sum_{j=1}^{4} d_{ie}, \quad G_{ij}^{\Omega z} = -\frac{\mu}{4}\sum_{j=1}^{4} e_{ie}$$

$$G_{ij}^{\Omega\Omega} = \frac{\mu}{6V^{(e)}}\sum_{j=1}^{4}(c_{ie}c_{je} + d_{ie}d_{je} + e_{ie}e_{je}), \quad W_i = -j\omega\mu H_{0x}V^{(e)},$$

$$X_i = -j\omega\mu H_{0y}V^{(e)}, \quad Y_i = -j\omega\mu H_{0z}V^{(e)}, \quad Z_i = \mu(c_{ie}H_{0x} + d_{ie}H_{0y} + e_{ie}H_{0z}).$$

Equation (4.147) means the potential equation focused on each element of the primary triangular element in Figure 4.17a.

Finally, to analyze the potential within each element using this type of matrix, it is necessary to convert its matrix into a matrix expressed by the entire node with reference to Eq. (4.147). That is, by rearranging Eq. (4.147) with attention paid to each node potential, an overall node equation can be obtained. This becomes possible since the relations $G_i = 0$ and $G_{di} = 0$ are held, that is, from the relationships of the following equations:

$$\sum_{i=1}^{ne} G_i = 0 \tag{4.148}$$

$$\sum_{i=1}^{ne} G_{di} = 0. \tag{4.149}$$

Therefore, by rearranging Eq. (4.147), denoted by each element for each node potential, an overall node equation can be obtained.

As an example, by calculating T based on the given analysis procedure, the eddy current density can be calculated from Eq. (4.97).

In this case, it should be noted that since the obtained eddy current density takes a constant value within the element, it becomes discontinuous on the boundary between the elements.

As a countermeasure, converting each element value obtained by applying the weighted average method to each node value is one of the effective solutions to this problem. Further, as a result of this analysis, the center of gravity of the element is shifted due to the method of element division, and a problem arises that a small error may occur in each node value.

4.3.8 Example of the Analytical Model

In this section, as a concrete example of the finite element method based on the analysis of the current vector potential, the topic of the eddy current absorber is introduced, associated with the development of high temperature heating implant in cancer thermotherapy [13, 14].

The thermal therapy described herein is a local thermal therapy in which an alternating magnetic field is irradiated to an implant buried in the body that is the object of heating. The applicator consists of ferrite cores. The eddy current absorber described herein is an important auxiliary means for achieving local hyperthermia with a high-temperature, heat-generating implant during ablation therapy.

As an analytical model, a heated object imitating a part of the body with a cubic agar phantom is taken up, as shown in Figure 4.18, and is installed between magnetic poles of the ferrite core. In this analysis, the total number of elements is 6000, and the total number of contact points is 1331.

Treatment with electromagnetic induction heating is used not only for the treatment of cancer but also for the treatment of shoulder pain, etc., in orthopedics, on a daily basis.

Regarding this type of thermal therapy, it becomes so difficult to regionally heat the intended treatment area. Hereafter, the methods for resolving these problems are clarified from the eddy current analysis data based on the finite element method, while introducing this reason. As an analytical model, a heated object imitating a part of the body with a cubic agar phantom is taken up, as shown in Figure 4.18, and is installed between magnetic poles of the ferrite core.

In addition, the phantom is made of an agar medium imitating the skin, muscle tissue, and the like, and has a cube shape with a side length of 20 cm. Here,

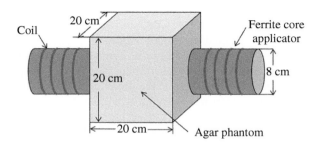

Figure 4.18 Configuration of magnetic field irradiator.

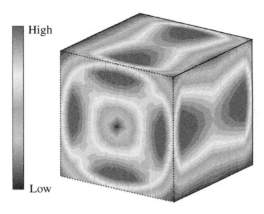

Figure 4.19 Three-dimensional analysis results of normalized average loss.

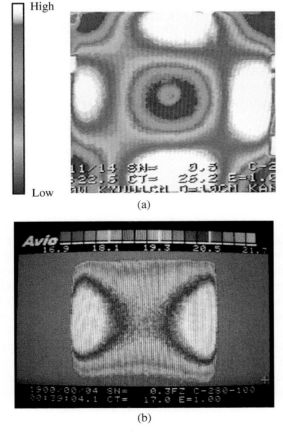

Figure 4.20 Surface temperature measurement by thermography. (a) Thermographic image after the heating experiment (front view of the phantom). (b) Thermographic image after the heating experiment (side view of the phantom).

we investigate the case when this phantom is placed between both magnetic poles made of cylindrical ferrite, and the implant is heated by applying an alternating magnetic field with a frequency of 4.0 MHz. We analyze the normalized average loss distribution during the irradiation by the magnetic field applied to the phantom to investigate the characteristics of the eddy current absorber.

The analytical result for the normalized average loss distribution calculated by the finite element method in the three-dimensional current vector potential method is shown in Figure 4.19 [15]. It should be noted that the normalized average loss distribution means the loss value normalized by the maximum loss value in the analytical region, and is an indicator of the calorific value in this analysis. In other words, the larger the normalized average loss value, the greater the heating amount of the phantom, that is, the higher the temperature is.

In the case of Figure 4.19, this is the result of analysis in a uniform phantom. Figure 4.20a,b shows thermographic views of the phantom surface and side faces obtained by experiments. From the analysis of the average loss distribution in Figure 4.19 and the thermographic image in Figure 4.20, we can see that a hot spot showing high temperature is generated in the peripheral portion of the phantom that is outside the center of the magnetic pole (the black portion in Figure 4.19 and the white portions in Figure 4.20). That is, we can find that the heat generation distribution due to the eddy current generated by the magnetic irradiation field strongly depends on the shape of the phantom and the electric constant distributions in the constituent medium of the phantom.

4.3.9 Unnecessary Current Absorber Analysis

Based on these investigations, next let us introduce the configuration method of an unnecessary eddy current absorber. Needless to say, in the unnecessary eddy current absorber configuration described here, we need to think about the appropriate configuration form according to its application. For example, the eddy current based on a large electric power of kW orders can occur, when considering the application to wireless electromagnetic induction type power feeding systems such as in cases of aiming at electric vehicle applications. On the other hand, the eddy current absorber in implant heating, introduced here, is normally used under the high-power magnetic radiations from 500 to 1000 W.

An analytical model of an unnecessary eddy current absorber is shown in Figures 4.21 and 4.22 [15]. This problem can be solved by introducing a three-dimensional vector current potential method and modeling the entire analysis region using the primary tetrahedral element, as described previously. In this analysis example, the total number of elements is 10 920 and the total number of nodes is 2430.

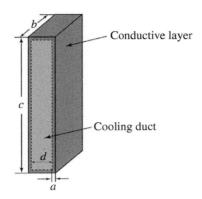

Figure 4.21 Configuration of the eddy current absorber.

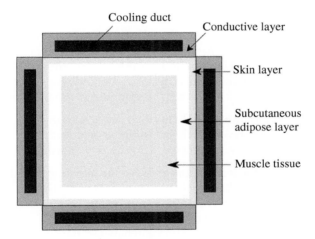

Figure 4.22 An analytical model when an eddy current absorber is loaded on a phantom.

The notations for each dimension of the unnecessary eddy current absorbers are given in Figure 4.21. The eddy current absorption bolus has a thickness $a = 0.3$ mm, a width $b = 20$ cm, and a length $c = 20$ cm. d (cm) is the lateral width of the bolus inside the cooling cavity, but in this analysis it is set to be constant 2 cm in all cases. First, for the case where an unnecessary current absorber is attached to the peripheral part of a cubic phantom having a predetermined conductivity, as shown in Table 4.3, the relationship between these dimensions and the optimum conductivity is investigated.

In this phantom model, the skin layer and subcutaneous fat layer are taken into consideration in order to approximate closely it as much as possible to

Table 4.3 Electrical constants of each part of human body model.

Material	Conductivity (S/m)	Relative permeability
Muscle tissue	0.62	1.0
Skin layer	0.62	1.0
Subcutaneous adipose layer	0.047	1.0
Conductive layer	Parameter	1.0
Cooling duct	1.0×10^{-3}	1.0

the actual human body. These material constants in each part of human body model are shown in Table 4.3.

Now let us investigate the relationship between the constituent dimensions of the eddy current absorbing bolus and its optimal conductivity value.

The results of the present theoretical analysis are exhibited in Figure 4.23a–c.

Here, the vertical axis on each graph represents the absorption bolus conductivity at optimal absorption characteristics, and the horizontal axis represents each dimension of the eddy current absorption bolus shown in Figure 4.22. Here, we focus on the following investigations about the eddy current absorber conductivities, which are required for the optimum absorbing characteristics.

(a) The case when the absorption bolus thickness a is changed,
(b) The case when the absorption bolus width b is changed,
(c) The case when the absorption bolus length c is changed.

Figure 4.23a shows the relationship between the dimensions of the eddy current absorption bolus thickness a and the optimal conductivity.

We can find that there is an inverse proportionality relationship between the absorption bolus thickness and the optimal conductivity.

Next, Figure 4.23b shows the relationship between the optimum conductivity of the absorption bolus and the width of the absorption bolus b. This figure confirms that the optimal absorption bolus conductivity is reduced by increasing the absorption bolus width, and it can be confirmed that there is an inverse proportionality relationship between these quantities.

Further, Figure 4.23c presents the result of the analysis of the relationship between the absorption bolus length c and the optimum conductivity of the absorption bolus. In this case, as the absorption bolus length c increases, its optimal conductivity increases, indicating that they are proportional to each other.

The actual eddy current absorbing devices are designed on the basis of theoretical analysis, and the data comparing measurement results and analytical

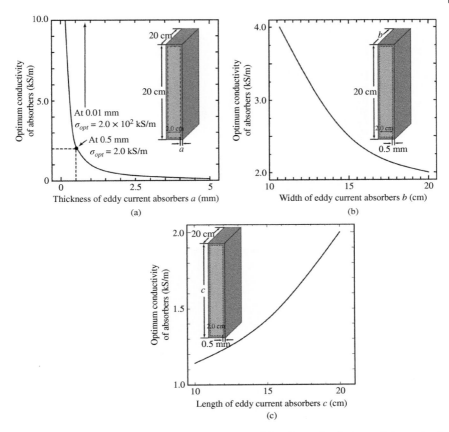

Figure 4.23 Relationship between each dimension of the absorption bolus and the optimal absorption bolus conductivity. (a) The case where the absorption bolus thickness a is taken as parameter. (b) The case where the absorption bolus width b is taken as parameter. (c) The case where the absorption bolus length c is taken as parameter.

results are shown in Figures 4.24 and 4.25, respectively [13, 14]. These figures show both the thermographic and analytical images when an implant that can generate heat by the magnetic field radiation into the phantom center region is embedded.

The nonuniform heat distribution, as shown in Figures 4.19 and 4.20, disappears, and it is visually confirmed that the eddy current absorber absorbs unnecessary eddy currents.

In recent years, the development of wireless electromagnetic induction resonance power feeding devices has attracted attention. Particularly in the

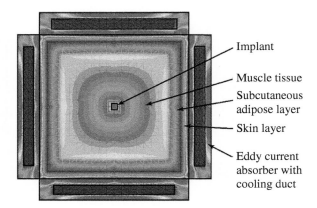

Figure 4.24 Heat distribution by absorption bolus using conductivity obtained by analysis.

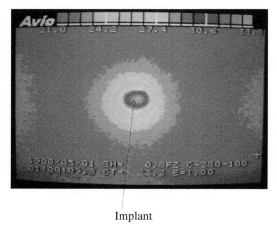

Implant

Figure 4.25 Measurement result of heat generation distribution using unnecessary current absorption bolus (thermography image, implant is buried in the center of the phantom).

application to automobiles, the surroundings are made of iron materials, and the radio wave disturbance due to the re-radiation field based on the generated eddy current is a concern. In this section, from the viewpoint of countermeasures against the thermal therapy problem, the concept of constructing an eddy current absorber was clarified by taking it as an example.

As anticipated beforehand, this eddy current absorber is basically based on Ohm's law. However, this type of eddy current absorption characteristics strongly depends on the shape of the object to be absorbed, and then the optimization of each dimension of the eddy current absorber and material constants is needed. Therefore, in order to ascertain the eddy current absorber characteristics, which optimize all the factors that constitute each part of the eddy current absorber, the FDTD computer analysis described is considered to be one of the effective means.

4.A Appendix

4.A.1 Appendix to Section 4.3.4 (1)

Volume coordinate expression of potential (Figure 4.A.1):

Let us explain again the derivation of the approximate current vector potential function $T^{(e)}$ in the primary tetrahedral elements of Figure 4.17a,b described in Section 4.3.4.

As pointed out in Section 4.3.4, when considering an arbitrary point $P(x, y, z)$ in this one primary tetrahedron element, this point $P(x, y, z)$ can further divide into four primary tetrahedron elements. Therefore, hereinafter, the total number of elements in the primary tetrahedral element is 4, that is, $k = 4$. If each of these four divided volumes is denoted by V_1, V_2, V_3, V_4, a volume coordinate variable can be given by the following equation:

$$\xi_k = \frac{V_k}{V^{(e)}} = \frac{1}{6V^{(e)}}(b_{ke} + c_{ke}x + d_{ke}y + e_{ke}z). \tag{4.A.1}$$

Here, $V^{(e)}$ represents the total volume of the primary tetrahedral element and $V_k(k = 1, 2, 3, 4)$ represents each volume which is divided into four. Note that the variable ξ_k means a variable that depends on the coordinates $P(x, y, z)$ of an arbitrary point in the primary tetrahedron element, as is apparent from Eq. (4.A.1). On the other hand, let us denote the potential $T^{(e)}(\xi_1, \xi_2, \xi_3, \xi_4)$ of an arbitrary point $P(x, y, z)$ in the primary tetrahedron element by the linear combination of ξ_k,

$$T^{(e)} = \alpha_1 \xi_1 + \alpha_2 \xi_2 + \alpha_3 \xi_3 + \alpha_4 \xi_4. \tag{4.A.2}$$

As can be seen from the volume relationship shown in definition expression (4.A.1), it turns out that ξ_k changes their values to $\xi_1 = 1, \xi_2 = \xi_3 = \xi_4 = 0$ when the point $P(x, y, z)$ moves to node 1 in primary tetrahedron element.

Using this property, when the potentials at the four nodes can be expressed as $T_{1e}, T_{2e}, T_{3e}, T_{4e}$, the unknown $\alpha_1, \alpha_2, \alpha_3, \alpha_4$ in Eq. (4.A.2) can be easily expressed.

Namely,

Potential of node 1: $T_{1e} = \alpha_1$
Potential of node 2: $T_{2e} = \alpha_2$
Potential of node 3: $T_{3e} = \alpha_3$
Potential of node 4: $T_{4e} = \alpha_4$.

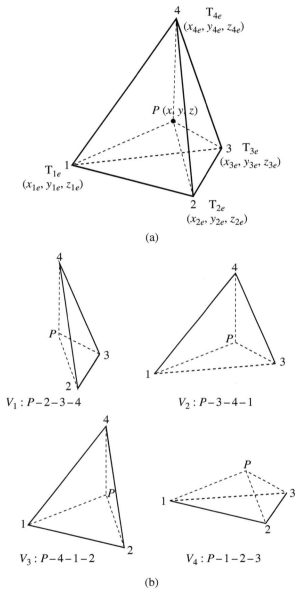

Figure 4.A.1 (a) Primary tetrahedron element. (b) Each tetrahedral element constructed by point P.

As a result, Eq. (4.A.2)

$$T^{(e)} = \begin{bmatrix} \xi_1 & \xi_2 & \xi_3 & \xi_4 \end{bmatrix} \begin{bmatrix} T_{1e} \\ T_{2e} \\ T_{3e} \\ T_{4e} \end{bmatrix} = \begin{bmatrix} \xi_1 & \xi_2 & \xi_3 & \xi_4 \end{bmatrix} T'_k \quad (k = 1, 2, 3, 4) \qquad (4.A.3)$$

where

$$T'_k = \begin{bmatrix} T_{1e} \\ T_{2e} \\ T_{3e} \\ T_{4e} \end{bmatrix},$$

T'_k is composed of the vector potentials $T_{1e}, T_{2e}, T_{3e}, T_{4e}$ at each node in the primary tetrahedral element. In this way, we find that the potential $T^{(e)}$ in the primary tetrahedral element can be related to the potential T'_k in each node.

Under these preparations, we next derive the relationship linking the node potential $T_{1e}, T_{2e}, T_{3e}, T_{4e}$ and the node coordinates (x_{ke}, y_{ke}, z_{ke}) in one primary tetrahedral element. The relationship between the arbitrary coordinate $P(x, y, z)$ inside the primary tetrahedron element and the volume coordinate variable ξ_k can be expressed by the following equation by representing the relation of the expression (4.A.1) in a matrix form:

$$\begin{bmatrix} \xi_1 & \xi_2 & \xi_3 & \xi_4 \end{bmatrix} = \frac{1}{6V^{(e)}} \begin{bmatrix} b_{1e} & b_{2e} & b_{3e} & b_{4e} \\ c_{1e} & c_{2e} & c_{3e} & c_{4e} \\ d_{1e} & d_{2e} & d_{3e} & d_{4e} \\ e_{1e} & e_{2e} & e_{3e} & e_{4e} \end{bmatrix} \begin{bmatrix} 1 & x & y & z \end{bmatrix}. \qquad (4.A.4)$$

Substituting Eq. (4.A.4) into Eq. (4.A.3), a matrix form of the potential $T^{(e)}$ at a point P in the primary tetrahedral element can be obtained:

$$T^{(e)} = \frac{1}{6V^{(e)}} \begin{bmatrix} 1 & x & y & z \end{bmatrix} \begin{bmatrix} b_{1e} & b_{2e} & b_{3e} & b_{4e} \\ c_{1e} & c_{2e} & c_{3e} & c_{4e} \\ d_{1e} & d_{2e} & d_{3e} & d_{4e} \\ e_{1e} & e_{2e} & e_{3e} & e_{4e} \end{bmatrix} \begin{bmatrix} T_{1e} \\ T_{2e} \\ T_{3e} \\ T_{4e} \end{bmatrix},$$

where $V^{(e)}$ is the total volume of the primary tetrahedral element and it is expressed by the following expression:

$$V^{(e)} = \frac{1}{6} \begin{vmatrix} 1 & x_{1e} & y_{1e} & z_{1e} \\ 1 & x_{2e} & y_{2e} & z_{2e} \\ 1 & x_{3e} & y_{3e} & z_{3e} \\ 1 & x_{4e} & y_{4e} & z_{4e} \end{vmatrix}.$$

Symbols (x_{ke}, y_{ke}, z_{ke}) are the node coordinates of the primary tetrahedral element.

Further, the following relation is held between the coordinate related to point $P(x, y, z)$ in the element and the volume coordinate variable $(\xi_1, \xi_2, \xi_3, \xi_4)$.

$$\begin{bmatrix} 1 & x & y & z \end{bmatrix} = \begin{bmatrix} 1 & x_{1e} & y_{1e} & z_{1e} \\ 1 & x_{2e} & y_{2e} & z_{2e} \\ 1 & x_{3e} & y_{3e} & z_{3e} \\ 1 & x_{4e} & y_{4e} & z_{4e} \end{bmatrix} \begin{bmatrix} \xi_1 & \xi_2 & \xi_3 & \xi_4 \end{bmatrix}. \tag{4.A.5}$$

From the relationships of these expressions (4.A.4) and (4.A.5), it is found that $b_{ke}, c_{ke}, d_{ke}, e_{ke}$ are associated with (x_{ke}, y_{ke}, z_{ke})

$$\begin{bmatrix} 1 & x_{1e} & y_{1e} & z_{1e} \\ 1 & x_{2e} & y_{2e} & z_{2e} \\ 1 & x_{3e} & y_{3e} & z_{3e} \\ 1 & x_{4e} & y_{4e} & z_{4e} \end{bmatrix}^{-1} = \frac{1}{6V^{(e)}} \begin{bmatrix} b_{1e} & b_{2e} & b_{3e} & b_{4e} \\ c_{1e} & c_{2e} & c_{3e} & c_{4e} \\ d_{1e} & d_{2e} & d_{3e} & d_{4e} \\ e_{1e} & e_{2e} & e_{3e} & e_{4e} \end{bmatrix}. \tag{4.A.6}$$

4.A.2 Appendix to Section 4.3.5 (1)

Derivation process of Eq. (4.130): y component in the first term of Eq. (4.110):

$$\left[-\frac{1}{\sigma} \iiint_V \nabla N_i \times \nabla \boldsymbol{T}\, dv \right]_y$$

$$= \iiint_V -\frac{1}{\sigma} \left\{ \frac{\partial N_i}{\partial z} \left(\frac{\partial T_z}{\partial y} - \frac{\partial T_y}{\partial z} \right) - \frac{\partial N_i}{\partial x} \left(\frac{\partial T_y}{\partial x} - \frac{\partial T_x}{\partial y} \right) \right\} dv$$

$$= \iiint_V -\frac{1}{\sigma} \left\{ \frac{e_{ie}}{6V^{(e)}} \left(\frac{\partial \sum_{j=1}^{4} N_j T_{zj}}{\partial y} - \frac{\partial \sum_{j=1}^{4} N_j T_{yj}}{\partial z} \right) \right.$$

$$\left. - \frac{c_{ie}}{6V^{(e)}} \left(\frac{\partial \sum_{j=1}^{4} N_j T_{yj}}{\partial x} - \frac{\partial \sum_{j=1}^{4} N_j T_{xj}}{\partial y} \right) \right\} dv$$

$$= \iiint_V -\frac{1}{\sigma} \sum_{j=1}^{4} \left\{ \frac{e_{ie}}{6V^{(e)}} \left(\frac{d_{je}}{6V^{(e)}} T_{zj} - \frac{e_{je}}{6V^{(e)}} T_{yj} \right) \right.$$

$$\left. - \frac{c_{ie}}{6V^{(e)}} \left(\frac{c_{je}}{6V^{(e)}} T_{yj} - \frac{d_{je}}{6V^{(e)}} T_{xj} \right) \right\} dv$$

$$= -\frac{1}{36V^{(e)2}\sigma} \sum_{j=1}^{4} (e_{ie} d_{je} T_{zj} - e_{ie} e_{je} T_{yj} - c_{ie} c_{je} T_{yj} + c_{ie} d_{je} T_{xj}) \times \iiint_V dv$$

$$= \frac{1}{36V^{(e)}\sigma} \sum_{j=1}^{4} \{-c_{ie}d_{je}T_{xj} + (c_{ie}c_{je} + e_{ie}e_{je})T_{yj} - e_{ie}d_{je}T_{zj}\}. \tag{4.130}$$

4.A.3 Appendix to Section 4.3.5 (2)

Derivation process of Eq. (4.131): z component in the first term of Eq. (4.110):

$$\left[-\frac{1}{\sigma}\iiint_V \nabla N_i \times \nabla T dv\right]_z$$

$$= \iiint_V -\frac{1}{\sigma}\left\{\frac{\partial N_i}{\partial x}\left(\frac{\partial T_x}{\partial z} - \frac{\partial T_z}{\partial x}\right) - \frac{\partial N_i}{\partial y}\left(\frac{\partial T_z}{\partial y} - \frac{\partial T_y}{\partial z}\right)\right\} dv$$

$$= \iiint_V -\frac{1}{\sigma}\left\{\frac{c_{ie}}{6V^{(e)}}\left(\frac{\partial \sum_{j=1}^{4} N_j T_{xj}}{\partial z} - \frac{\partial \sum_{j=1}^{4} N_j T_{zj}}{\partial x}\right)\right.$$

$$\left. -\frac{d_{ie}}{6V^{(e)}}\left(\frac{\partial \sum_{j=1}^{4} N_j T_{zj}}{\partial y} - \frac{\partial \sum_{j=1}^{4} N_j T_{yj}}{\partial z}\right)\right\} dv$$

$$= \iiint_V -\frac{1}{\sigma}\sum_{j=1}^{4}\left\{\frac{c_{ie}}{6V^{(e)}}\left(\frac{e_{je}}{6V^{(e)}}T_{xj} - \frac{c_{je}}{6V^{(e)}}T_{zj}\right)\right.$$

$$\left. -\frac{d_{ie}}{6V^{(e)}}\left(\frac{d_{je}}{6V^{(e)}}T_{zj} - \frac{e_{je}}{6V^{(e)}}T_{yj}\right)\right\} dv$$

$$= -\frac{1}{36V^{(e)2}\sigma}\sum_{j=1}^{4}(c_{ie}e_{je}T_{xj} - c_{ie}c_{je}T_{zj} - d_{ie}d_{je}T_{zj} + d_{ie}e_{je}T_{yj}) \times \iiint_V dv$$

$$= \frac{1}{36V^{(e)}\sigma}\sum_{j=1}^{4}\{-c_{ie}e_{je}T_{xj} - d_{ie}e_{je}T_{yj} + (c_{ie}c_{je} + d_{ie}d_{je})T_{zj}\}. \tag{4.131}$$

References

1 Yee, K.S. (1966). Numerical solution of initial boundary value problem involving Maxwell's equations in isotropic media. *IEEE Trans. Antennas Propag.* 14 (3): 302–307.
2 Taflove, A. (1998). *Advances in Computational Electrodynamics The Finite-Difference Time-Domain Method*. Artech House Publishers.
3 Mur, G. (1981). Absorbing boundary conditions for the finite-difference approximation of the time domain electromagnetic field equations. *IEEE Trans. Electromagn. Compat.* 23: 377–382.
4 Berenger, J.P. (1994). A perfectly matched layer for the absorption of electromagnetic waves. *J. Comput. Phys.* 114: 184–200.

5 Amano, M. and Kotsuka, Y. (2003). A method of effective use of ferrite for microwave absorber. *IEEE Trans. Microwave Theory Tech.* 51 (1): 238–245.

6 Courant, R. (1943). Variational method for the solution of problems of equilibrium and variation. *Bull. Am. Math. Soc.* 4 (9): 1–23.

7 Turner, M.J., Clough, R.W., Martin, H.C., and Tops, L.P. (1956). Stiffness and deflection analysis of complex structures. *J. Aeronaut. Sci.* 23 (9): 805–823.

8 Melosh, R.J. (1963). Basis of derivation of matrices for different stiffness method. *AIAA J.* 1: 204–223.

9 Clough, R.W. (1960). The finite element method in plane stress analysis. In: *Proceedings of. 2nd Conference on Electronic Computation, ASCE, Pittsbutgh, PA*, 344–378.

10 Kotsuka, Y., Hankui, E., Hashimoito, M., and Miura, M. (2000). Development of double-electrode applicator for localized thermal therapy. *IEEE Trans. Microwave Theory Tech.* 48 (11): 1906–1908.

11 Kotsuka, Y. and Tanaka, T. (1999). Method of improving EM-field distribution in a small room with an RF radiator. *IEEE Trans. Electromagn. Compat.* 41 (1): 22–29.

12 Isono, I. (1994). Master's thesis. Tokai University Graduate School, pp. 46–68.

13 Kotsuka, Y., Hankui, E., and Sigematsu, Y. (1996). Development of ferrite core applicator system for deep-induction hyperthermia. *IEEE Trans. Microwave Theory Tech.* 44 (10): 1803–1810.

14 Kotsuka, Y., Watanabe, M., Hosoi, M., and Isono, I. (2000). Development of inductive regional heating system for breast hyperthermia. *IEEE Trans. Microwave Theory Tech.* 48 (11): 1807–1814.

15 Kotsuka, Y., Kayahara, H., Murano, M. et al. (2009). Local inductive heating method using novel high-temperature implant for thermal treatment of luminal organs. *IEEE Trans. Microwave Theory Tech.* 57 (10): 2574–2580.

5

Fundamental EM-Wave Absorber Materials

Needless to say, in the design of EM-wave absorbers, their characteristics depend largely on the characteristics of the electromagnetic (EM)-wave absorbent material itself. In this chapter, in order to obtain basic knowledge of carbon materials and ferrite materials, which are typical EM-wave-absorbing materials, their respective characteristics are examined, especially from the viewpoint of molecular and crystal structures. The Jaumann absorber in early stage of the EM-wave absorber was composed of overlaying the material dispersed carbonyl iron to rubber, the resistive sheet, and the dielectric plate, as described in Chapter 1. In this way, since the absorber was invented, carbon-based materials have been mainly used as EM-wave-absorbing materials. On the other hand, copper–zinc-based soft ferrite was invented in 1930 in Japan. However, the nature of this kind of soft ferrite as an EM-wave absorber was unknown for a long time. Magnetic materials typified by ferrite, metallic resistors, and carbon materials used as resistive films are applied to EM-wave absorbers, and they have changed little to this day.

In this chapter, these materials are outlined only from the limited viewpoint of their molecular and crystal structures in order to acquire the basic knowledge of EM-wave-absorbing materials, focusing on carbon materials and ferrite materials. First, Section 5.1 examines carbon materials used as absorbers and attenuator element materials. In Section 5.2, to clarify what kind of ferrite property among many magnetic materials is used as the absorber materials, our attention focuses on the ferrite crystal structure.

5.1 Carbon Graphite

In this section, the crystal structure of the carbon material is outlined first.

Carbon-based materials are the mainstream of the dielectric-type EM-wave absorbers, resistive EM-wave absorbers, and attenuator element materials. Generally, a solid is largely divided into (i) a crystal and (ii) an amorphous material. (i) A "complete crystal" has such a structure that the construction

Electromagnetic Wave Absorbers: Detailed Theories and Applications, First Edition. Youji Kotsuka.
© 2019 John Wiley & Sons, Inc. Published 2019 by John Wiley & Sons, Inc.

of atoms is regularly arranged, and the same structural units are arranged three-dimensionally and indefinitely. However, for actual crystals, "single crystals" surrounded by several finite planes or "polycrystalline," which are collections of small crystals, exist.

A typical example of a single crystal is diamond. As examples of a polycrystalline, metals that are routinely used can be considered. (ii) "The amorphous material" is typified by glass, in which the atomic arrangement of a solid is disturbed and has no regularity. In addition, coal, charcoal, soot, animal charcoal, gas carbon, and so on are known as a group of allotropes that do not show a clear crystalline state in the form of amorphous carbon in a carbon material. Carbon-based materials were the mainstream of materials absorbing EM-waves, and attenuator element materials.

As an example of an EM-wave absorber using these materials, an absorber made by combining a carbon material and a rubber, an absorber comprising a carbon fiber and a rubber, and the like was proposed. Further, a lightweight EM-wave absorber made of graphite and polystyrene foam, with excellent electromagnetic-wave absorption characteristics, is also in practical use. Graphite, known as the pencil core material, is a carbon allotrope and is also used as an electromagnetic-wave absorber material.

Carbon usually has four covalent bonding hands from the center of a tetrahedron, but this bonding hand may stay in the plane in some cases. Regarding this explanation, although details are omitted, this depends on the combination of the trajectories of carbon electrons. As shown in Figure 5.1a, the basic planar structure of carbon atoms in graphite is represented as a regular plane hexagon where six carbon atoms have bond hands. This is called a benzene nucleus structure. Graphite is composed of carbon atoms spread two-dimensionally in the form of a network with a benzene nucleus structure, and they are further stacked up, as shown in Figure 5.1b.

Depending on how the hexagonal network plane is oriented, it becomes possible to realize a carbon material having various characteristics. Concerning this orientation form, the non-orientation, axial orientation, point orientation, plane orientation, etc. can exist. For example, the carbon fiber has concentric-shape crystalline structures which are oriented in the axial direction, as shown in Figure 5.1c,d. Here, we have mainly discussed the graphite as a carbon material that can be manufactured industrially and can easily control electrical losses.

By the way, there also exist carbon materials used for EM-wave absorbers which are produced in nature. For instance, Figure 5.2 shows the measured data to compare the EM-wave absorber characteristics according to the types of charcoal routinely used. In the present case, three kinds of charcoal, such as black charcoal, white charcoal produced from the ubame oak tree (Japanese tree name), and bamboo charcoal, are adopted to examine the EM-wave absorber characteristics of carbon materials existing in nature. The black

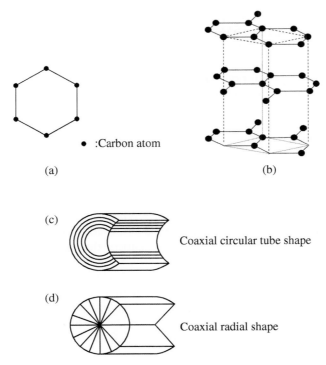

Figure 5.1 Crystal structure of carbon graphite. (a) Planar structure of the carbon atom. (b) Graphite consists of planar structures of the carbon atom which are stacked up. (c) Concentric crystalline structure oriented in the axial direction of the cylinder. (d) Crystalline structure oriented in the radial direction of the cylinder.

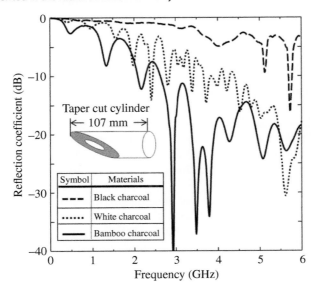

Figure 5.2 EM-wave-absorbing characteristics of the charcoal in nature.

charcoal is carbonized at a carbonization temperature of 400–700 °C and is called black charcoal because its surface is black.

White charcoal is made by burning again almost perfect charcoal at a temperature of about 1300°C. It is called white charcoal because white ash remains on the charcoal surface after burning. Bamboo charcoal is made of bamboo. This charcoal has a major feature that it can be burned at low temperature (400°C), medium temperatures (600 – 700°C), and high temperature (1000°C), using the same charcoal grill kiln depending on various applications.

The measurement sample as the EM-wave absorber described herein is shaped into a cylinder by mixing each charcoal powder with an epoxy resin material. The reason why the sample is produced in a cylindrical shape is to measure the EM-wave absorption characteristics using a coaxial waveguide. The cylindrical charcoal pipe is loaded so as to close contact with the shorted terminal end of the 20D coaxial waveguide. The total length of the sample is 10.4 cm. In this case, assuming a pyramidal absorber, the reflection coefficient measurements are performed by obliquely cutting the incident surface side of the cylindrical pipe into a taper shape.

From these data, it can be estimated that bamboo charcoal has the best EM-wave absorption characteristic among them. In this way, we can learn the complexity of carbon materials in that the electric conductivity varies greatly depending on the properties of the wood used and the firing temperature, etc., even though it seems to be the same material at first sight.

5.2 Ferrite

In this section, among various magnetic materials, what ferrite constituting a ferrite absorber is positioned, and what kinds of properties exist are summarized.

5.2.1 Soft Magnetic Material

Magnetic materials are classified into hard and soft magnetic materials by means of the coercive force, which is a measure of the magnetic hardness, as shown in Figure 5.3. A ferromagnetic material having a large coercive force is called a "hard magnetic material," and it is represented by a permanent magnet. In contrast, a magnetic material with a small coercive value is called a "soft magnetic material," and, for example, it is used as an iron core in a transformer.

The distinction between soft and hard magnetic materials is not rigorously clear, but a magnetic material with a coercive force value above 100 [Oe] ($\fallingdotseq$8000 A/m) is usually defined as a hard magnetic material. The nomenclature of "hard" and "soft" is said to come from the fact that hardened steel is

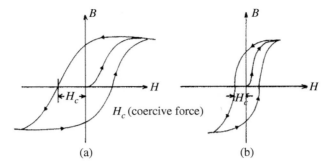

Figure 5.3 Magnetic coercive force. (a) Hard magnetic material (H_c: large values). (b) Soft magnetic material (H_c: small values).

mechanically strong and the coercive force is high, whereas the steel softened by tempering decreases the coercive force. Ferrite for EM-wave absorbers, widely used in general, belong to the "soft magnetic materials." For this reason, even if it is magnetized, it does not become a permanent magnet.

5.2.2 Spinel-type Magnetic Oxide

Ferrites are commonly called an iron oxide having a molecular formula of $MO \cdot Fe_2O_3$ [1].

Here, the symbol M is usually a divalent metal ion and could be Mn^{2+}, Zn^{2+}, Fe^{2+}, Co^{2+}, Ni^{2+}, Cu^{2+}, Mg^{2+}, and the like; also, the ferrites can be composed of a mixture of two or more ions of these kinds. Here, firstly, we examine the crystal structure of a general oxide, and then we describe that the ferrite used in the EM-wave absorber is a "magnetic oxide having a spinel-type crystal structure."

5.2.2.1 Crystal Structure of Oxide

In general, the oxide represented by MO_x ($x = n/2$) is a compound in which a metal ion M^{n+} is bound to an oxygen ion O^{2-}. In many cases, the value of n is 2 or 3, and in rare cases it can contain four ions or one.

When this oxide is present as a solid, a single-molecule MO_x does not simply exist in the form of agglutination due to the cohesive force, but rather is based on the Coulomb force acting between the positive and negative ions (charges $+ne$ and $-2e$) of M^{n+} and O^{2-}, respectively.

This type of crystal is called an ionic crystal. In the oxide crystal, the O^{2-} ion radius is 1.32 Å and the metal ion M^{n+} radius is about 0.6–0.8 Å. This radius is smaller than for the O^{2-}. Therefore, as shown in Figure 5.4, when the spheres are depicted when considering the radial dimensions of these two kinds of large and small ions (O^{2-} and M^{n+}, respectively), the large O^{2-} ion can be represented so as to be in contact with each other.

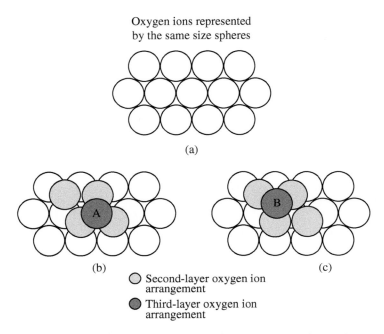

Oxygen ions represented
by the same size spheres

(a)

(b)   (c)

◯ Second-layer oxygen ion
    arrangement
● Third-layer oxygen ion
    arrangement

Figure 5.4 Atomic layout of the crystal's closest packing structure. (a) Oxygen ion arrangement of the first layer. (b) Oxygen ion arrangement of the third layer. (c) Different oxygen ion arrangement of the third layer.

These figures indicate cases in which oxygen atoms drawn in a spherical shape are arranged in close contact with each other. That is, the second layer of gray color is stacked on top of the first layer of the white circles, and the third layer of dark gray (only one sphere is shown) is stacked on top of the second layer of gray circles. In this case, as a method of stacking oxygen atoms in the third layer, there are two ways to put them at either A or B, as shown in Figure 5.4b,c, respectively.

In the case of Figure 5.4b, the oxygen atom of the third layer is located just above the oxygen atom of the first layer. This structure is named as a hexagonal close-packed structure. Also, in (c), the oxygen atoms in the third layer are located directly above the valleys of the arrangement of atoms in the first layer. Such a crystal structure is called a *cubic* close-packed structure (face-centered cubic).

Now, paying attention to the first and second layers in Figure 5.4b, in a position where the first layer is seen from point A, the gaps between the three oxygen ions of the second layer and one oxygen ion in the first layer exist. Also, paying attention to the position where the first layer should be seen from point B in Figure 5.4c, there is a gap consisting of three oxygen ions in the second layer and three ions in the first layer. As can be seen from the fact just mentioned, ferrite properties are dominated by the type of oxygen ions or metal ions of ferrite entering positions A and B. This point is discussed in detail in the next section.

5.2.2.2 Crystal Structure of Ferrite

As described, generally, ferrite is an oxide in a ceramic ferrimagnet represented by the chemical formula $MO \cdot Fe_2O_3$. Here, M is a divalent metal ion such as Mn^{2+}, Zn^{2+}, Fe^{2+}, Co^{2+}, Ni^{2+}, Cu^{2+}, Mg^{2+}, and the like, or their mixtures.

Since the oxygen ions themselves do not have magnetism, the magnetic properties of ferrite are determined by how metal ions (Fe, M) enter positions A and B (that is, the relationship of the respective magnetic moments). As a unit cell of the crystal structure of ferrite, A is composed of gaps into which 8 ions can be introduced, and B is composed of gaps into which 16 ions can enter. On the other hand, the number of oxygen ions is 32 in one crystal structure unit of ferrite.

That is, considering the ratio of the number of gaps in which ions of A and B can enter, and the number of oxygen ions, this is 1 : 2 : 4.

On the other hand, in the molecular formula of ferrite, the ratio of the ion numbers of M, Fe, and O is 1, 2, and 4, respectively. Therefore, since M is 1, Fe is 2, and O is 4 in the molecular formula of ferrite, it seems reasonable if M is positioned at point A and Fe is positioned at point B.

However, in practice, Neel discovered that the ions do not necessarily take such an arrangement, and he demonstrated the case where one of the two Fe atoms belongs to A and the other Fe belongs to B, and M occupies the remaining vacant position B. This case is called "inverse spinel," and it represents a ferromagnetic nature because of the magnetic moment M^{2+} that exists in position B. On the other hand, there are groups in which all Fe take position B and M takes position A. This is what is referred to as "positive spinel," because the magnetic moment at each position A and B is fully offset, and this represents a paramagnetic nature. In addition, "spinel" is named after the fact that ferrite has the crystal structure of natural ore spinel ($MgAl_2O_4$) and the same isomorphic structure. For this reason, ferrite is called the crystal of the spinel structure.

Table 5.1 Main compositions of M^{2+}.

Ni Zn
Ni Cu Zn
Ni Mg Zn
Ni Zn Co
Ni Cu Zn Co
Mn Zn
Cu Zn

After all, from the viewpoint of the crystal structure, the usual ferrite absorbers have a structure of inverse spinel and are said to be a "polycrystalline composite ferrite," which is composed of a mixture of two or more kinds of ions, like M^{2+}. The main compositions of M^{2+} are shown in Table 5.1.

5.3 Hexagonal Ferrite

In the previous section, a general polycrystalline ferrite was described, which was used in the EM-wave absorber in relatively lower microwave frequency regions. Recently, from the development requirements of the EM-wave absorber and millimeter-wave devices in the millimeter-wave regions, hexagonal ferrite was introduced which can maintain magnetic properties in the millimeter-wave band. This is called a magnetoplumbite-type oxide, and ions having $Ba^{2+}, Sr^{2+}, Pb^{2+}$, and the like with large radii, are included along with the Fe^{3+} ion and divalent metal ion M^{2+} of the ferrite. For this reason, the crystal structure forms a hexagonal crystal structure, as if the position of O^{2+} in ferrite was replaced by these ions. In this case, depending on the structure of the layer, including the third ion Ba^{2+}, etc., and the ratio of the spinel layer, the hexagonal crystals are classified into M, W, Y, and Z type, and so on. In this way, various magnetic compounds are produced. This hexagonal ferrite is characterized by a large anisotropic magnetic field of the material, so that the natural resonant frequency becomes high and the configuration of the EM-wave absorber in the millimeter-wave range becomes possible. That is, in hexagonal ferrite, an EM-wave absorber in a high-frequency band is realized by having an "anisotropic magnetic field" in the material itself. This fact can be illustrated from the standpoint of analogy by magnetizing a normal sintered ferrite with a DC magnetic field in the microwave frequency band. In short, because of the natural resonance frequency shift to higher frequency region, the absorbing frequency can be realized even at microwave frequency band. For reference, the EM-wave absorption characteristic is shown in Figure 5.5 when an ordinary polycrystalline ferrite disk is loaded at the end of a coaxial short terminal waveguide and a DC magnetic field is applied. In this case, the DC magnetic field is applied to the waveguide axis direction by a coil wound around the coaxial waveguide to realize the ferrite anisotropy.

This DC field application generates an anisotropic characteristic in the ferrite, and the magnetic permeability characteristic, that is, the resonance frequency of the permeability, moves to the higher frequency region. Consequently, it becomes easy to understand the actions described if we consider a hexagonal ferrite – a magnetic material, composed similarly to this basic principle. The ferrite used here is a Ni–Zn-based polycrystalline ferrite [2].

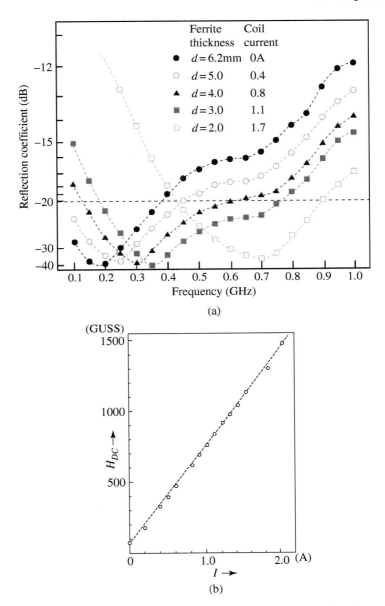

(a)

(b)

Figure 5.5 Electromagnetic-wave absorption characteristics in the case when the anisotropy is realized by ordinary polycrystalline ferrite. (a) Changes in EM-wave absorption characteristics when a DC field is applied perpendicular to the ferrite surface. Here, however, the coil current being wound around the outer periphery of the coaxial waveguide is expressed as a parameter. (b) Relationship between the current flowing through the coil wound around the coaxial waveguide outer conductor and the DC magnetic field generated by the current.

References

1 Soohoo, R.F. (1985). *Microwave Magnetics*. Harper & Row Publishers Inc.
2 Kotsuka, Y. and Yamazaki, H. (2000). Fundamental investigation on a weakly magnetized absorber. *IEEE Trans. Electromagn. Compat.* 42 (2): 116–124.

6

Theory of Special Mediums

In this section, chiral, magnetized ferrite, and metamaterial media are taken up as special media related to the EM-wave absorber and attenuator, and these theoretical treatments of EM fields are described in detail.

Section 6.1 focuses on introducing the physical meaning of chiral media, because the method of electromagnetic-field analysis treats extremely complicated expressions. Therefore, only the analytical procedures are described here. As an application example of a chiral medium to the EM-wave absorber, an absorber in the low-frequency range (1–6 GHz) of the microwave frequency band is introduced.

In the first half of Section 6.2, in order to derive the tensor permeability that is the theoretical foundation of magnetized ferrite, fundamental matters such as the angular momentum of electrons and precession movement are explained together with the introduction of physical images. On the basis of these fundamental matters, the motion equation of the magnetization is derived, and the procedure for deriving the tensor permeability and the circularly polarized permeability in the case where the magnetic medium is lossless or lossy is described in detail. The problems of electromagnetic-wave propagation in magnetized ferrite media, as for the phenomena of the rectangular coordinate system, are described in many books. For this reason, Section 6.3 takes up the derivation method of the formula with respect to the theoretical equations in the cylindrical coordinate system, along with the methods of electromagnetic-wave analysis in a circular waveguide. Further, the analytical method of coaxial waveguides is explained in detail. The reason for this is that knowledge about the problem with loss of ferrite, such as in the case of electromagnetic-wave absorption and attenuation characteristics, can be deduced relatively easily from the investigation of propagation characteristics inside the coaxial waveguide in place of the free-space treatment. In Section 6.4, the concept of a left-handed metamaterial is investigated from the viewpoints of its implementation in EM-wave absorbers. After explaining basic metamaterial knowledge such as the development process of metamaterials,

Electromagnetic Wave Absorbers: Detailed Theories and Applications, First Edition. Youji Kotsuka.
© 2019 John Wiley & Sons, Inc. Published 2019 by John Wiley & Sons, Inc.

negative permittivity and permeability, and negative refractive index media, configuration examples of the absorber are described.

6.1 Chiral Medium

It is estimated that the chiral medium is one candidate for an EM-wave absorber material, particularly in the millimeter-wave region. The actual situations of the chiral medium have not been fully elucidated from the standpoint of the EM-wave absorber. In the future, along with the development of new technologies for the millimeter-wave and terahertz frequency fields, the chiral material is expected to be potentially used for progressing to new applications, as with metamaterials.

In the high-frequency range, ferrite cannot be used essentially because of Snoek's limit. In contrast, a chiral medium can be used in the high-frequency region, because it can be essentially composed of the dielectric material. At present, the chiral material is not developed enough to be sufficiently available. Accordingly, as an introduction to it, only the physical meanings and the theoretical analysis that assume the EM-wave absorber applications are described in this section.

A chiral medium includes chiral substances in a dielectric medium, and it has properties similar to those of an optically active medium. A chiral object is defined as an object that cannot hold geometric congruence, even if its mirror image and the object itself translate in parallel or rotate.

As familiar examples of chiral objects, the relationships of counterclockwise- and clockwise-rotated helical conductors and a pair of gloves are shown in Figure 6.1.

Generally, an electromagnetic wave that is incident on some material acts on atoms or molecules in the material and induces a polarization phenomenon.

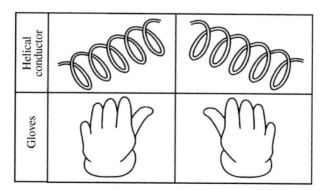

Figure 6.1 Examples of chiral medium structures.

Thereby, this phenomenon generates a secondary electric field or a secondary magnetic field in the material. For materials with structures having normal symmetry centers, however, these secondary electromagnetic waves are canceled out. In contrast, when the molecules and crystal structures constituting the material have a special nature, the secondary effect appears. Therefore, the electromagnetic-wave analysis problem in this type of material must treat electromagnetic fields by taking this secondary effect into account. A chiral medium is a medium in which incident electromagnetic waves are converted into circularly polarized waves under the action of electromagnetic fields generated by this secondary effect.

The physical meaning of generation of this circularly polarized wave is shown in Figure 6.2.

Figure 6.2b,c shows the cases where electromagnetic waves are incident on the conductor coils placed horizontally or vertically. The conductor coil is assumed here to be a right-handed coil.

As shown in Figure 6.2b, let us consider a case where the right-wound conductor coil is placed perpendicular to the incident electric field E and the magnetic field H is directed to the coil axis direction. In this case, an electric field E_H corresponding to the electromotive force due to the incident magnetic field H is generated along the coil, based on Faraday's law. At this time, as shown in the circle of the dotted line in the figure, the x-direction component E_{Hx} of E_H

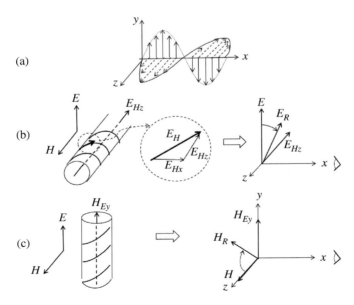

Figure 6.2 Behaviors of the incident wave in a helical conductor coil.

is canceled out, because opposite electromotive forces occur mutually at every pitch of the coil. As a result, only the E_{Hz} component of the coil axial direction is generated as a secondary electric field. When this secondary electric field E_{Hz} and the incident wave electric field E are synthesized, as shown in the figure with an arrow, as the resultant field viewed from the x-axis, a clockwise electric field E_R can be generated.

Next, when the right-wound conductor coil, as shown in Figure 6.2c, is placed vertically against H – that is, it is put horizontally along with the incident electric field E – a magnetic field H_{Ey} is generated in the same direction as the incident wave electric field E. As a result, the synthesized magnetic field H_R by the incident magnetic field H and the secondary magnetic field H_{Ey} rotates in the clockwise direction as viewed from the x-axis, as shown in Figure 6.2c. In this case also, the left-wound coil generates a magnetic field $-H_{Ey}$ in the direction opposite to E.

In the given explanation, we focused on the relation between the incident electric field and the direction of the coil. However, even when considering the incident magnetic field, the magnetic field rotates to the right with the right-wound coil and to the left with the left-wound coil, in the same way as described previously.

In these examples, the incident electric field and the magnetic field were horizontal or perpendicular, respectively, to the conductor coil. In general, however, when the incident electric field or the magnetic field are inclined with respect to the conductor coil axis (actually, the axis of the molecule), the same rotation occurs. This can be understood by considering the parallel and perpendicular components of the incident electric field or the magnetic field with respect to the coil axis.

6.1.1 Electromagnetic Fields in Chiral Medium

Let us now consider the case when a plane electromagnetic wave E_i with arbitrary polarization is incident at an angle θ_i from the dielectric medium I, assuming free space to the chiral medium II, as shown in Figure 6.2. Here, ε, μ are the permittivity and permeability of each medium, respectively, and β is a constant called chirality admittance of medium II, which can take positive and negative values [1–3].

In Figure 6.3, $E_{i\parallel}, E_{r\parallel}$, and $E_{i\perp}, E_{r\perp}$ mean the horizontal polarization components and the vertical polarization components of the electric field vector of the incident wave, respectively.

Generally, the constitutive equations of electromagnetic fields in a chiral medium are expressed as follows [1, 2]:

$$D = \varepsilon E + j\beta B, \tag{6.1}$$

$$H = j\beta E + (1/\mu)B. \tag{6.2}$$

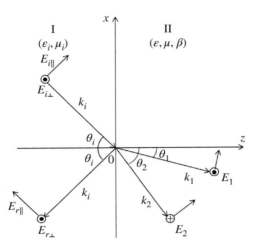

Figure 6.3 Reflected and transmitted waves at the chiral medium interface.

We now consider the case when there is no EM-wave source in the area under consideration; in this case, Maxwell's equations are given by the following expressions:

$$\nabla \times \boldsymbol{E} = j\omega \boldsymbol{B} \tag{6.3}$$

$$\nabla \times \boldsymbol{H} = -j\omega \boldsymbol{D}. \tag{6.4}$$

By deriving the Helmholtz equation including the chiral medium from Eqs. (6.1)–(6.4), the following equation is obtained:

$$\nabla^2 \boldsymbol{C} + 2\omega\mu\beta\nabla \times \boldsymbol{C} + k^2 \boldsymbol{C} = 0. \tag{6.5}$$

Here, $\boldsymbol{C}$ represents $\boldsymbol{E}, \boldsymbol{H}, \boldsymbol{D}$, and $\boldsymbol{B}$, corresponding to the electric field, the magnetic field, the electric flux density, and the magnetic flux density, respectively, in a unified equation (6.5).

In addition, here the wave number for a general medium, including a chiral medium, is represented by $k = \omega\sqrt{\varepsilon\mu}$. If $\beta = 0$, it represents the wave number of a normal medium.

Now, the electromagnetic field which satisfies Eq. (6.5) is represented by the superposition of a right-handed circularly polarized wave $\boldsymbol{E}_1$ and a left-handed circularly polarized wave $\boldsymbol{E}_2$ with different phase velocities. The wave numbers k_1 and k_2 of the respective circularly polarized waves in the present case are given by the following equations:

$$k_1 = \omega\mu\beta + \sqrt{\omega^2\mu^2\beta^2 + k^2} \tag{6.6}$$

$$k_2 = -\omega\mu\beta + \sqrt{\omega^2\mu^2\beta^2 + k^2} \tag{6.7}$$

Here, the wave number in the medium I is,

$$k_i = \omega\sqrt{\varepsilon_i\mu_i},$$

When using these expressions, it is found that the phase velocities of each circularly polarized wave are expressed as ω/k_1 and ω/k_2, respectively, and different values depending on the nature of the chirality admittance β are given. That is, in the case of $\beta > 0$, the phase velocity of the clockwise circularly polarized wave is later than the phase velocity of the counterclockwise circularly polarized wave, and when $\beta < 0$, the phase velocity is reversed. The axis of the propagation direction of the electromagnetic field in each region is defined by Snell's law, and the following relationship holds:

$$k_i \sin\theta_i = k_1 \sin\theta_1 = k_2 \sin\theta_2. \tag{6.8}$$

Here, θ_1 and θ_2 are the refractive angle of the right-handed circular polarization wave and the left-handed circular polarization wave, respectively.

6.1.2 Electromagnetic-Field Reflection by Chiral Medium

The analytical representations of reflection and transmission in chiral media become very complicated. For this reason, this section outlines only the method of finding the reflection coefficient of the electromagnetic-wave absorber like a flowchart. As for this kind of detailed method of analysis, refer to the literature [5]. Now, as shown in Figure 6.4, let us consider a plane incident EM wave with the incident angle θ_0 in the free space with an arbitrary polarization incident on the chiral medium which is attached to a conductor plate.

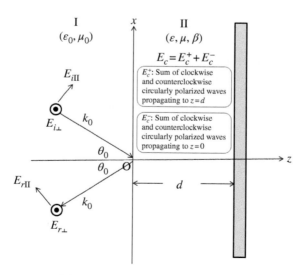

Figure 6.4 Construction of the chiral medium EM-wave absorber.

In the following, the medium constant of medium I in Figure 6.1 is replaced by the vacuum medium constant, and the indication related to this medium is represented by the suffix 0. In Figure 6.1, $E_{i\|}, E_{r\|}$, and $E_{i\perp}, E_{r\perp}$ represent the horizontal polarization and the vertical polarization component of the electric field vectors of the incident wave, respectively.

Therefore, Eq. (6.8) is rewritten by the following equation:

$$k_0 \sin \theta_0 = k_1 \sin \theta_1 = k_2 \sin \theta_2. \tag{6.9}$$

Now, the incident plane wave in medium I has an arbitrary polarization plane and the electric field is given by the following equation. To avoid complex descriptions of the present explanations, the magnetic field equations are omitted in the discussion (see Appendix 6.A.1).

$$\left. \begin{array}{l} E_i = E_{0i} e^{jk_0(z \cos \theta_0 - x \sin \theta_0)} \\ E_{0i} = E_{i\perp} e_y + E_{i\|}(\cos \theta_0 e_x + \sin \theta_0 e_z) \end{array} \right\}. \tag{6.10}$$

Here, (e_x, e_y, e_z) presents unit vectors corresponding to the x, y, and z axis, respectively. Also, the subscript $\perp$ and $\|$ represent the vertical and horizontal polarization components of the electric field, respectively. Notice here that the incident power is normalized to one in the following analysis. That is, the incident power is expressed by the following equation:

$$|E_{i\perp}|^2 + |E_{i\|}|^2 = 1. \tag{6.11}$$

Similarly, the electric field of the reflected wave in medium I can be expressed by the following equation:

$$E_r = E_{0r} e^{-jk_0(z \cos \theta_0 + x \sin \theta_0)}, \tag{6.12}$$

where

$$E_{0r} = E_{r\perp} e_y + E_{r\|}(\cos \theta_0 e_x - \sin \theta_0 e_z).$$

Next, the electric fields in the chiral medium II are given by the relations of the following expressions:

$E_c^+ = $ (The sum of the clockwise circularly polarized wave and the counterclockwise circularly polarized wave, propagating in the direction of $z = d$).

$E_c^- = $ (The sum of the clockwise circularly polarized wave and the counterclockwise circularly polarized wave, propagating in the direction of $z = 0$).

In these expressions, the subscripts + and − represent the directions of waves propagating to the right and left in medium II, respectively.

That is, the total electric field in the chiral medium II, which is aimed at the final expressions of the electric field, is given by

$$E_c = E_c^+ + E_c^-, \tag{6.13}$$

where

$$E_c^+ = E_1^+ e^{jk_1(z \cos \theta_1 - x \sin \theta_1)} + E_2^+ e^{jk_1(z \cos \theta_1 - x \sin \theta_1)}, \tag{6.14a}$$

$$E_c^- = E_1^- e^{-jk_1(z\cos\theta_1 + x\sin\theta_1)} + E_2^- \cdot e^{-jk_2(z\cos\theta_2 + x\sin\theta_2)}, \tag{6.14b}$$

and

$$E_1^+ = E_1^+(\cos\theta_1 e_x + \sin\theta_1 e_z + je_y) \tag{6.15}$$

$$E_2^+ = E_2^+(\cos\theta_2 e_x + \sin\theta_2 e_z - je_y) \tag{6.16}$$

$$E_1^- = E_1^-(\cos\theta_1 e_x - \sin\theta_1 e_z + je_y) \tag{6.17}$$

$$E_2^- = E_2^-(\cos\theta_2 e_x - \sin\theta_2 e_z - je_y). \tag{6.18}$$

Here the suffixes 1 and 2 indicate the clockwise circularly polarized wave and the counterclockwise circularly polarized wave, respectively.

The magnetic field vectors in medium I and medium II can be found from Maxwell's equations in the same manner as in the case of the electric field (see Appendix 6.A.1).

Next, the reflected power from the chiral medium necessary for analyzing the EM-wave absorber has to be derived. To derive a wave reflected from the chiral medium, by applying the boundary conditions at the boundary surface $z = 0$ and $z = d$, the following Equation (6.19) is obtained. Here, because the analytical derivation of the present expressions becomes too complicated, the derivation processes of the detailed equation are omitted and only analytical procedures are formally described.

(a) First, the boundary condition is applied with respect to the incident wave and the reflected wave in free space and the electromagnetic field in the chiral medium in the $z = 0$ plane. The electric field in the chiral medium is represented by Equation (6.13). The equation of the magnetic field can be obtained in the same way.

(b) Next, let us consider the boundary condition on the plane $z = d$. Regarding the electric field, the electric field at the boundary surface is defined by Equation (6.13) with $z = d$.

$$\begin{bmatrix} E_{i\perp} \\ E_{i\|} \\ E_{r\perp} \\ E_{r\|} \end{bmatrix} = [A] \begin{bmatrix} E_1^+ \\ E_2^+ \\ E_1^- \\ E_2^- \end{bmatrix} \tag{6.19}$$

$$\begin{bmatrix} E_1^+ \\ E_2^+ \\ E_1^- \\ E_2^- \end{bmatrix} = \frac{ZK}{4S_1 S_2} \begin{bmatrix} S_2 \\ S_1 \\ S_1^2 S_2 \\ S_1 S_2^2 \end{bmatrix} = [B] \tag{6.20}$$

where, $S_1 = e^{jk_1 d \cos\theta_1}$, $S_2 = e^{jk_2 d \cos\theta_2}$, $Z = (\mu/\varepsilon)^{1/2}[1 + (\mu/\varepsilon)\beta^2]^{-1/2}$ (value of chiral medium), K [A/m] is current density at the boundary surface $z = d$. Since there is no need in the following description, the concrete matrix notation of [A] is omitted.

In Eqs. (6.19) and (6.20), [A] and [B] represent the matrices of 4×4 and 4×1, respectively. The elements of each matrix are given as the functions of such variables as the angle of the incident wave, each material constant in the mediums, and their thicknesses. From Eqs. (6.19) and (6.20), the electric field components of the incident wave and the reflected wave are given by the following equation using the medium constants:

$$\begin{bmatrix} E_{i\perp} \\ E_{i\parallel} \\ E_{r\perp} \\ E_{r\parallel} \end{bmatrix} = [A][B] = [T]. \tag{6.21}$$

Here, $[T]$ is the 4×1 matrix.

From Eq. (6.21), $E_{r\perp}$ and $E_{r\parallel}$ are represented by the following expressions using $E_{i\perp}$ and $E_{i\parallel}$.

$$E_{r\perp} = \frac{T_2}{T_1} E_{i\perp} + \frac{T_2}{T_3} E_{i\parallel} \tag{6.22}$$

$$E_{r\parallel} = \frac{T_4}{T_1} E_{i\perp} + \frac{T_4}{T_3} E_{i\parallel} \tag{6.23}$$

Here, T_1, T_2 etc. represent the elements of the matrix [T]. Further, the reflected power P_r is expressed by the following equation:

$$P_r = P_{r\perp} + P_{r\parallel}. \tag{6.24}$$

Here, $P_{r\perp}$ and $P_{r\parallel}$ denote that the powers are composed of the perpendicular and parallel components of reflected waves, respectively.

They are expressed by the following equations, using each electric field component, given by Eqs. (6.22) and (6.23):

$$P_{r\perp} = |E_{r\perp}|^2 \tag{6.25}$$

$$P_{r\parallel} = |E_{r\parallel}|^2. \tag{6.26}$$

From these equations, the reflection coefficient can be calculated.

The above-described is the outline of the analysis of a single-layer absorber model, as shown in Figure 6.4. When using a single-layer chiral medium absorber, only the reflection coefficient of the order of 3–4 dB can be obtained at the frequency of the microwave band, as shown in Figure 6.5.

Incidentally, if ferrite powder is added to the chiral medium, it has been found that EM-wave absorption characteristics of 20 dB or less can be obtained even in the microwave band [6].

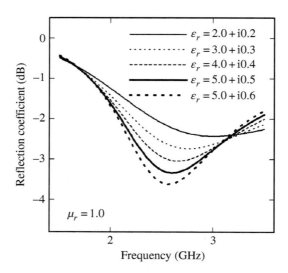

Figure 6.5 EM-wave absorption characteristics in microwave band in a single-layer chiral medium.

Next, let us introduce an EM-wave absorber composed of multi-chiral mediums, as shown in Figure 6.6. Regarding the problems of the influence of chiral concentration rates on reflection, they have been already investigated [4]. Hence, the EM-wave absorber characteristics from the viewpoint of changing the dielectric material constant of each layer composed of a chiral medium are introduced here. These investigations are conducted by the theoretical

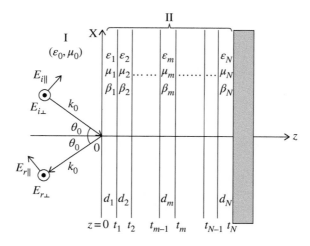

Figure 6.6 Geometry of the EM-wave absorber composed of multi-chiral mediums.

formula which was derived for the multilayered chiral mediums on the basis of the previous theoretical analysis in a single-layer chiral medium [4, 5].

The point to note with this type of numerical analysis is that the chirality admittance β cannot take an arbitrary value. The chirality admittance β_m of the mth layer is limited by the following equation [3]:

$$|\beta_m| \leq \sqrt{\frac{\varepsilon_m}{\mu_m}}.$$

Here, ε_m and μ_m represent the dielectric and permeability constants of the mth layer.

In the entire present medium, the permeability value μ_m is 1.0 and β is set to $\beta = 7.5 \times 10^{-4} [S]$.

The notations, such as CBA in Figure 6.7, mean that each permittivity is stacked in the order of the CBA values shown in Table 6.1. A good absorption is around 2.7 GHz in the case of Figure 6.7.

In the present case, the thickness of each chiral medium part is set to an almost constant value close to $\lambda / 6$, which is the absorbing frequency wavelength that we want.

Also, Figure 6.8 shows the EM-wave absorption characteristics in the case of 5 layers and 10 layers with the permittivity combinations shown in Table 6.1 [6].

As for the chiral medium, it seems difficult to realize the EM-wave absorption characteristics in a low-frequency region, such as the microwave frequency

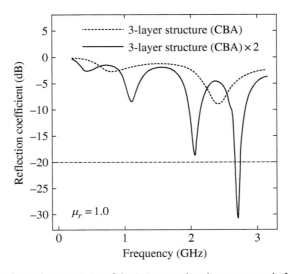

Figure 6.7 Absorbing characteristics of the EM-wave absorber composed of 3- and 6-layer chiral mediums.

Table 6.1 Dielectric constant of chiral media.

Medium	Relative permittivity
A	$2.0 + i0.2$
B	$3.0 + i0.3$
C	$4.0 + i0.4$

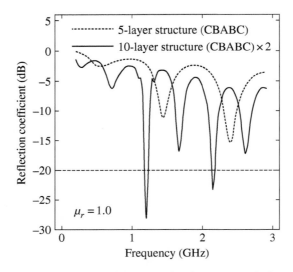

Figure 6.8 Absorbing characteristics of EM-wave absorber composed of 5- and 10-layer chiral mediums.

band, because of the nature of the chiral medium. However, as a result of the analysis based on the above mentioned theory, EM-wave absorption characteristics can be improved even in the low-frequency range of 1–5 GHz microwave band by implementing the multilayered chiral medium with a few kinds of permittivity.

6.2 Theory of Magnetized Ferrite

A typical magnetic absorber was principally realized by a ferrite material.

As mentioned in Chapter 5, ferrite is usually used as EM-wave-absorbing materials in the states of sintered ferrite, rubber ferrite, plastic ferrite, or the like. When applying a static magnetic field to these ferrite materials, electrically controllable interesting characteristics can be available by an

EM-wave absorber or attenuator. Moreover, by focusing on the method of applying a static magnetic field to the ferrite, we can deepen our understanding of the relationships between the permeability characteristics and the absorbing characteristics of the EM-wave absorber or physical properties related to spin behavior.

6.2.1 Foundation of Equation of Magnetization Motion

In this section, let us first investigate the equation of magnetization motion in both lossless and lossy cases, and then analyze the propagation characteristics for magnetization ferrite, particularly in the cylindrical coordinate system [7, 8].

Now, as for ferrite, it has a 10^{14}-fold resistivity compared to metal, and the dielectric constant is 10^{15} or more, which means a high-insulating magnetic material. Generally, it is characterized as a kind of metal oxide having the molecular formula $MO \cdot Fe_2O_3$.

Here, M represents divalent metals such as nickel, manganese, zinc, iron, cadmium, magnesium, and so on, as described in Chapter 5. The magnetic properties of ferrite are governed primarily by the magnetic dipole moment associated with the electron rotation. In general, as is well known, the cause that produces a magnetic moment can be divided into two phenomena.

One is an electron self-rotation and the other is electron circulation movement in orbit around the nucleus. By the way, in ferromagnetics such as ferrite, the proportion contributing to magnetic moment generation due to the self-rotation of an electron is overwhelmingly large. Based on this fact, we shall take into account only self-rotation of electrons hereafter.

As shown in Figure 6.9, when one electron is in self-rotation (spinning), this phenomenon is considered equivalent to the current flowing against the direction of a self-rotation of an electron in Figure 6.9a. The equivalent transformation processes between one electron rotation and a magnetic dipole are shown in Figure 6.9a–c. The intention of replacing a self-rotating electron with an electron circulating in orbit is to introduce the concept of a material point. In short, we can associate the current problems with the concept of momentum in mechanics.

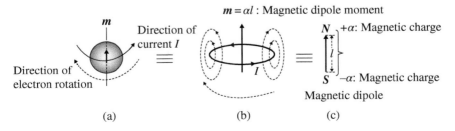

Figure 6.9 Electron rotation and magnetic dipole.

At the N and S poles of the micro-magnets, giving the magnetic charges of $+\alpha$ and $-\alpha$ in Figure 6.9c, the magnetic moment m is defined as follows:

$$m = \alpha l, \tag{6.27}$$

where l is the distance between the magnetic poles S and N.

Now that this micro-magnet (electron) is placed in a uniform magnetic field H, the torque T, given by the following equation, acts on this magnetic moment:

$$T = m \times H. \tag{6.28}$$

On the other hand, since the electron spins with a constant mass, it is considered that it has an angular momentum P from the mechanics viewpoint. The angular momentum P is defined as the vector product $r \times G$ represented by the radial vector r that extends from the origin to the point mass, as shown in Figure 6.10a. Here, G represents the momentum of the material point in the mechanics. In this consideration, the material point can be regarded approximately as if located in origin. This is because the radius value r is very small. Hence, the direction of the angular momentum P is opposite to the direction of the magnetic moment. Since the magnitude of the angular momentum P is proportional to the magnetic moment m, the following relation is satisfied:

$$\frac{m}{P} = \gamma. \tag{6.29}$$

Here, γ is called the gyromagnetic ratio, and has a negative sign, as is clear from the fact that P and m are antiparallel as shown in Figure 6.10b. Using this

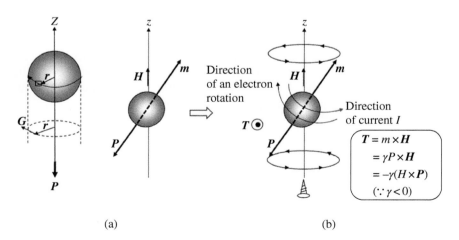

(a) (b)

Figure 6.10 Image of precession motion. (a) Explanation of angular momentum. (b) Precession movement.

relationship, Eq. (6.28) is

$$T = m \times H = \gamma P \times H. \tag{6.30}$$

In the present case, since the torque T is applied to the magnetic moment m, as a result, the magnetic moment m comes to do the precession motion regarding to the H direction axis, as shown in Figure 6.10c. Now, the equation of the magnetization motion is given by the following equation, based on the definition that the changing rate of the angular momentum is equal to the torque T:

$$\frac{dP}{dt} = T. \tag{6.31}$$

In Eq. (6.31), by substituting Eqs. (6.28) and (6.29),

$$\frac{dm}{dt} = \gamma m \times H. \tag{6.32}$$

As shown in Figure 6.10, when the DC magnetic field H is applied to one rotating electron, from the relation of the right-hand side $\gamma m \times H$ in Eq. (6.32), the tip of the m continues the precession movement (note $\gamma < 0$) in the present lossless case. By the way, only one electron in the magnetic crystal has been considered so far. However, magnetic crystals, such as ferrite, contain countless electrons, for some of which the magnetic moment m faces in opposite directions.

Hence, when considering the entire crystal medium, we need to take the sum of the individual magnetic moment **m** vectors in Eq. (6.32). Accordingly, the sum of the magnetic momenta in a unit volume can be expressed using the relation $M = \sum m$ in place of m in Eq. (6.32).

Then,

$$\frac{dM}{dt} = \gamma M \times H. \tag{6.33}$$

This expression gives the equation of the magnetization motion in the lossless case.

The present case, when M is in the precessional motion relative to the DC magnetic field H, is called free precession motion. The angular frequency (angular velocity) ω_0 in this case is also called the natural precession frequency of a magnetic dipole in a uniform magnetic field H_0.

This relation is given by

$$\omega_0 = \gamma H_0.$$

Further, when a small alternating magnetic field is added to the DC magnetic field to overlap, the magnetic moment performs a forced precession motion.

Equation (6.33) was the case when precession occurs in the lossless magnetic material. But when the magnetic material has the loss, the equations of the magnetization motion with the loss term were proposed by Landau–Lifshitz and Bloch–Bloembergen.

These equations are

Landau–Lifshitz formula:

$$\frac{dM}{dt} = \gamma(M \times H) - \lambda \left\{ \frac{(H \cdot M)M}{M^2} - H \right\} \tag{6.34}$$

Where, λ is a damping factor possessing the dimensions of a frequency and it can be also called the relaxation frequency. It has the dimension of the inverse of a relaxation time.

Bloch–Bloembergen formula:

$$\left. \begin{array}{l} \left(\dfrac{dM}{dt}\right)_{x,y} = \gamma(M \times H)_{x,y} - \dfrac{M_{x,y}}{T_2} \\[3mm] \left(\dfrac{dM}{dt}\right)_{z} = \gamma(M \times H)_{z} - \dfrac{M_{z}}{T_1} \end{array} \right\} \tag{6.35}$$

Here, T_1 represents the spin–lattice relaxation time and T_2 denotes a loss associated with any process that disturbs or opposes precession motion. If this kind of loss exists, while the precession motion radius of the magnetization vector shown in Figure 6.10c is swirling, its radius gradually decays. As a result, all the magnetization inside the crystal is aligned ultimately in the direction of the DC magnetic field, and the saturation magnetization appears as a whole.

6.2.2 Tensor Susceptibility

6.2.2.1 Lossless Medium Case

First, let us consider the tensor susceptibility without taking into account crystal losses. This tensor susceptibility is derived from the equation of the magnetization motion in Eq. (6.33) in the lossless case. In Eqs. (6.32) and (6.33), we assumed H is only a uniform magnetic field or a DC magnetic field. However, in practical applications, such as microwave isolators or circulators, etc., the DC magnetic field H that is superimposed by the RF magnetic field (the microwave field), has to be taken into consideration as a total magnetic field H. In this respect, the case where the DC magnetic field and the r-f magnetic field preserve, the orthogonal relation becomes particularly important.

Now, let us represent the total magnetic field H in this case by the following equation:

$$H = H_i + he^{j\omega t}. \tag{6.36}$$

Here, H_i is the vector sum of the total DC magnetic fields in a magnetic material, taking into account any time-independent internal field. The second term means a high-frequency magnetic field with the amplitude h and the angular frequency ω of the r-f field (basically, the microwave frequency field). We assume, in the present case, the relation $H_i \gg h$ (the small signal condition). In accordance with Eq. (6.36), the magnetization M is rewritten by the equation related to the steady DC magnetization vector M_0 and the r-f magnetization vector m_f (the microwave magnetization vector):

$$M = M_0 + m_f e^{j\omega t}. \tag{6.37}$$

Since assuming the relations of $H_i \gg h$ in Eq.(6.36), $M_0 \gg m_f$ (M_0 is the maximum DC magnetization) is held in (6.37). Here, let us substitute Eqs. (6.36) and (6.37) for H and M into Eq. (6.33) and expand the M in terms of the exponential components of the time. Since all terms of higher order than the power of $e^{j\omega t}$ in the small quantities h and m_f can be neglected first, the DC equation is represented as

$$\gamma(M_0 \times H_i) = 0. \tag{6.38}$$

This development is usually called the small signal approximation. This DC equation is based on the fact that M_0 and H_i are in the same direction.

Next, when omitting the higher order $e^{j\omega t}$ term and equalizing the first-order $e^{j\omega t}$ term, the following a-c equation can be obtained:

$$j\omega m_f = \gamma(M_0 \times h \div m_f \times H_i). \tag{6.39}$$

From this equation, the equation for the susceptibility of the medium can be derived, as well as the condition for the resonance evaluation. Here, in order to simplify the treatment of the equations given later, let us assume that the medium has an infinite length and an anisotropy based on the crystal lattice, that is, magnetostriction is zero, and only the external static magnetic field H_0 contributes to H_i. Hence, $H_i = H_0$ in the following.

First, multiplying the H_0 from the right to Eq. (6.39),

$$j\omega m_f \times H_0 = \gamma(M_0 \times h) \times H_0 \div \gamma(m_f \times H_i) \times H_0. \tag{6.40}$$

And then, this equation is expanded using the triple vector product formula $A \times (B \times C) = (A \cdot C)B - (A \cdot B)C$.

Hence,

$$j\omega m_f \times H_0 = \gamma(H_0 \cdot M_0)h - \gamma(H_0 \cdot h)M_0 + \gamma(H_0 \cdot m_f)H_0 - \gamma(H_0 \cdot H_0)m_f. \tag{6.41}$$

And, also, taking the scalar product of Eq. (6.39) and H_0 and using the expression (6.38), the following expression can be derived:

$$j\omega m_f \cdot H_0 = \gamma(M_0 \times h) \cdot H_0 + \gamma(m_f \times H_0) \cdot H_0$$
$$= h \cdot \gamma(H_0 \times M_0) + m_f \cdot \gamma(H_0 \times H_0) = 0. \tag{6.42}$$

Here, because the relations $H_0 \times M_0 = 0$ from Eq. (6.38) and $H_0 \times H_0 = 0$ are held, it is found that $m_f \cdot H_0 = 0$. This means that r-f magnetization m_f is perpendicular to the applied field H_0. Consequently, the third term of Eq. (6.41) may disappear, and then we can solve it for m_f:

$$m_f = \frac{1}{\omega_0{}^2 - \omega^2} \{ j\omega\gamma (M_0 \times h) + \gamma^2 (H_0 \cdot M_0)h - \gamma^2 (H_0 \cdot h)M \}. \qquad (6.43)$$

Here, $\omega_0 = \gamma H_0$ (angular velocity during free precession) (see Appendix 6.A.2).

As is apparent from this equation, m_f has a singularity in the case $\omega = \omega_0 = \gamma H_0$.

This relation denotes the resonance condition. Now, we represent each coordinate component of m_f by

$$m_f = im_x + jm_y + km_z.$$

The tensor of magnetization can be derived from Eq. (6.43) in the following manner.

First, let us assume that both H_0 and M_0 values are oriented in the same direction as the z axis of the Cartesian coordinate system, and the applied r-f magnetic field can be expressed in the following component equation:

$$h = ih_x + jh_y + kh_z. \qquad (6.44)$$

By substituting Eq. (6.44) and the relations of $H_0 = kH_0$ and $M_0 = kM_0$ into Eq. (6.43), and by expanding it and then equating each x, y, z component, the following expressions can be obtained:

$$\left. \begin{aligned} m_x &= \frac{\gamma M_0}{\omega_0{}^2 - \omega^2}(\omega_0 h_x - j\omega h_y) \\ m_y &= \frac{\gamma M_0}{\omega_0{}^2 - \omega^2}(j\omega h_x + \omega_0 h_x) \\ m_z &= 0 \end{aligned} \right\}. \qquad (6.45)$$

Now, since the r-f magnetization vector m_f can be associated with the r-f magnetic field h, the relation between these quantities is denoted as a magnetic susceptibility χ,

$$m_f = \chi h. \qquad (6.46a)$$

When Eq. in (6.46) is expressed in a matrix form,

$$\begin{bmatrix} m_x \\ m_y \\ m_z \end{bmatrix} = \chi \begin{bmatrix} h_x \\ h_y \\ h_z \end{bmatrix} = \begin{bmatrix} \chi_{xx} & \chi_{xy} & \chi_{xz} \\ \chi_{yx} & \chi_{yy} & \chi_{yz} \\ \chi_{zx} & \chi_{zy} & \chi_{zz} \end{bmatrix} \begin{bmatrix} h_x \\ h_y \\ h_z \end{bmatrix}. \qquad (6.46b)$$

From the relationship of Eq. (6.45) and (6.46), the following expressions are obtained:

$$\left. \begin{aligned} m_x &= \chi_{xx}h_x + \chi_{xy}h_y = \frac{\gamma M_0 \omega_0}{\omega_0{}^2 - \omega^2}h_x - j\omega\frac{\gamma M_0}{\omega_0{}^2 - \omega^2}h_y \\ m_y &= \chi_{yx}h_x + \chi_{yy}h_y = \frac{\gamma M_0}{\omega_0{}^2 - \omega^2}j\omega h_x + \frac{\gamma M_0 \omega_0}{\omega_0{}^2 - \omega^2}h_y \\ m_z &= 0 \end{aligned} \right\}. \qquad (6.47)$$

Comparing Eq. (6.45) and (6.46), the tensor form of the magnetic susceptibility is given by the following tensor form:

$$\chi = \begin{bmatrix} \chi_{xx} & \chi_{xy} & 0 \\ \chi_{yx} & \chi_{yy} & 0 \\ 0 & 0 & 0 \end{bmatrix}.$$

Here,

$$\chi_{xx} = \frac{\omega_0 \omega_M}{\omega_0^2 - \omega^2}(= \chi_{yy}) \tag{6.48}$$

$$\chi_{yx} = \frac{j\omega \omega_M}{\omega_0^2 - \omega^2}(= -\chi_{xy}). \tag{6.49}$$

Here, $\omega_M = \gamma M_0$ (M_0 is the saturation magnetization oriented in the H_0 direction).

Now, in practical applications, such as ferrite at the microwave frequency, we often encounter the permeability expression, rather than the susceptibility. Generally, the magnetic flux density B and the magnetic field H in the scalar expression are in the relation $B = \mu_0(H + M) = \mu_0(1 + \chi)H = \mu H$, when the intensity of magnetization is defined by M/μ_0. Under these conditions, the tensor permeability can be derived in the following manner. As the first step, each component of the magnetic flux density needs to be expressed in the form of simultaneous equations using Eq. (6.47). Then, let us take here the procedure for denoting these simultaneous equations with the matrix form. As a result, the relationship of the following equation is finally obtained for the magnetic flux density $B = (B_x, B_y, B_z)$:

$$\begin{bmatrix} B_x \\ B_y \\ B_z \end{bmatrix} = \mu_0 \begin{bmatrix} 1 + \chi_{xx} & \chi_{xy} & 0 \\ \chi_{yx} & 1 + \chi_{yy} & 0 \\ 0 & 0 & 1 \end{bmatrix} \begin{bmatrix} h_x \\ h_y \\ h_z \end{bmatrix} = \mu_0 \begin{bmatrix} \mu & -j\kappa & 0 \\ j\kappa & \mu & 0 \\ 0 & 0 & 1 \end{bmatrix} \begin{bmatrix} h_x \\ h_y \\ h_z \end{bmatrix}.$$

Here, $\mu = 1 + \chi_{xx}$, $j\kappa = -\chi_{xy}$.

This tensor notation of permeability $\bar{\mu}$ (conventionally called the Polder tensor) is

$$\bar{\mu} = \mu_0 \begin{bmatrix} \mu & -j\kappa & 0 \\ j\kappa & \mu & 0 \\ 0 & 0 & 1 \end{bmatrix}. \tag{6.50}$$

Finally, when trying to elucidate the physical properties of ferrite, this is conventionally considered to be a research theme in a field close to physics, rather than in the electromagnetic-wave engineering. On the basis of this, we have to keep in mind that CGS units, which were used as the unit system for studying the physical properties of ferrite, are often introduced instead of MKS units. In fact, ferrite company data, etc., are often presented in the units of CGS, and the units such as Gauss meters, etc., are also used in measuring instruments. In the case of CGS units, the saturation magnetization is expressed in Gauss and the magnetic field in oersted. Therefore, we should pay attention to the selection of CGS or MKS units in the actual application of the theory. Namely, when a unit in Eq. (6.46a) is given by CGS unit (Gaussian units) $\omega_M = \gamma M_0$ can be represented as $\omega_M = \gamma(4\pi M_0)$. Then $4\pi M_0$ is denoted in gauss and magnetic field with $\omega_0 = \gamma H_0$ in oersteds. The gyromagnetic ratio γ is $\gamma = 1.76 \times 10^7$ rad/s Oe in CGS unit for $g = 2$. When this value is used, the angular frequency $\omega = 2\pi f$ is expressed in radian per second and f is the frequency in Hertz per second. As for microwave applications, it is useful to express the processional frequency in megahertz per second. For this case, $\gamma' = 2.8$MHz/sOe, where $f(\text{MHz}) = 2.8H(\text{Oe}) = [\gamma/(2\pi \times 10^6)H]$.

6.2.2.2 Loss Medium Case

Throughout the previous analysis, any losses associated with the motion of a dipole in an actual ferromagnetic medium were neglected. In this section, the focus is on the theory of tensor magnetic susceptibility in a lossy medium, where the precession of M is gradually attenuated by internal loss. In this case, it is necessary to adopt the magnetization motion equation, which has an additional loss term (the damping term).

As in the magnetization motion equation, which has the loss term, two basic equations, described in the previous section, are widely used. One is the Landau–Lifshitz (L–L) form, expressed in Eq. (6.34), and the other is the Bloch–Bloembergen (B–B) form in Eq. (6.35). The L–L form can determine the susceptibility tensor from a relatively simple analysis, and it can be applied to a wide range of applications in the microwave device fields. On the other hand, the B–B form is an effective expression when discussing the relaxation process.

In this section, we deal with the L–L form to derive the tensor magnetic susceptibility including the loss. Referring to Eq. (6.34) again,

$$\frac{d\mathbf{M}}{dt} = \gamma(\mathbf{M} \times \mathbf{H}) - \lambda \left\{ \frac{(\mathbf{H} \cdot \mathbf{M})\mathbf{M}}{M^2} - \mathbf{H} \right\} \quad (6.34).$$

As stated previously in Equation (6.34), λ is a damping factor possessing the dimensions of the frequency, and it can also be called the relaxation frequency. It has the dimension of the inverse relaxation time. By applying the vector formula $\mathbf{A} \times (\mathbf{B} \times \mathbf{C}) = \mathbf{B}(\mathbf{A} \cdot \mathbf{C}) - \mathbf{C}(\mathbf{A} \cdot \mathbf{B})$ to the second term on the right-hand side of Eq. (6.34),

$$\frac{d\mathbf{M}}{dt} = \gamma(\mathbf{M} \times \mathbf{H}) - \frac{\alpha\gamma}{|M|}\{\mathbf{M} \times (\mathbf{M} \times \mathbf{H})\}. \quad (6.51)$$

Here, only for the sake of convenience, let us substitute the relation $\alpha\gamma/|M| = \lambda/|M^2|$.

The DC magnetic field H_r, when μ''_+ becomes maximum value, is called a resonance magnetic field. In addition, the width between the DC magnetic fields, when the value of the loss term μ''_+ takes a half value at the maximum value, is called the half-width of the magnetic resonance. The maximum value of μ'_+ is given at the point of $H_r + \Delta h/2$, and its minimum value is given at the point $H_r - \Delta h/2$. To summarize the behavior of positive and negative circularly polarized waves, the relative permeability takes a large value near the point of magnetic resonance. And, for certain values of H_0, it can also have negative values. In this way, the relative permeability of positive and negative circularly polarized waves takes large values near the point of magnetic resonance, as seen from the figure. And μ'_+ exhibit negative values at certain values of H_0. Finally, we note in this figure that we assumed that H_0 means only a pure DC magnetic field inside the magnetic material.

6.3 MW-Propagation of Circular Waveguide with Ferrite

Microwave propagation problems in the case of the mounting of ferrite in a rectangular waveguide were studied in many literature sources. For this reason, let us consider the circular waveguide structures, such as a circular waveguide or a coaxial waveguide, in which ferrite is mounted fully or partially. In general, when a microwave propagates in a magnetized ferrite anisotropic medium, the phenomenon of rotation of the polarization wave plane arises and phase (angle) changes are caused.

Based on the M.L. Kales theory [9], EM-field equations of these kinds are taken up here, particularly from the viewpoints of EM-wave absorption and attenuation.

6.3.1 Derivation of Fundamental Equations

In this section, we derive the fundamental field equations for the microwave propagation in a ferrite anisotropic medium in a cylindrical coordinate system [9]. Let us first consider the case when ferrite is loaded into a circular waveguide, as shown in Figure 6.12. In this figure, the DC magnetic field H_{DC} is applied to the direction of the waveguide axis z.

Therefore, the ferrite is assumed to be uniformly magnetized by H_{DC} in the waveguide axis direction. In addition, the microwave magnetic field is assumed to be sufficiently small (signal small conditions), compared with the magnitude of the DC magnetic field. Since we consider the case of applying a DC magnetic field to ferrite, the magnetic permeability $\hat{\mu}$ is represented by the Polder tensor, but the permittivity ε is assumed to be isotropic.

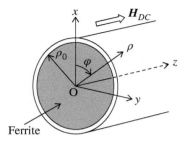

Figure 6.12 Ferrite-loaded circular waveguide.

When denoting the time factor by $e^{j\omega t}$, the Maxwell equations are

$$\nabla \times H = j\omega D$$
$$\nabla \times E = -j\omega B$$
$$\nabla \cdot D = 0$$
$$\nabla \cdot B = 0, \tag{6.68}$$

where

$$D = \varepsilon E$$
$$B = \hat{\mu} H = \begin{bmatrix} \mu & -j\mu' & 0 \\ j\mu' & \mu & 0 \\ 0 & 0 & \mu_z \end{bmatrix} \begin{bmatrix} H_r \\ H_\theta \\ H_z \end{bmatrix}. \tag{6.69}$$

Assuming that i_z is a unit vector in the axial direction z, the subscript t indicates those associated with the x- and y-axis direction components, and the propagation constant in the guide axis direction is given by γ. At this point, the following relationships of Eq. (6.70) are satisfied between B and H:

$$B = \mu H_t + j\mu' i_z \times H_t + \mu_z H_z i_z. \tag{6.70}$$

From the Maxwell equations,

$$\nabla_t \times E + \gamma i_z \times E = -j\omega B \tag{6.71}$$
$$\nabla_t \times H + \gamma i_z \times H = j\omega \varepsilon E. \tag{6.72}$$

And,

$$\nabla_t \times E = \nabla_t \times E_t + \nabla_t E_z \times i_z \tag{6.73}$$
$$\nabla_t \times H = \nabla_t \times H_t + \nabla_t H_z \times i_z. \tag{6.74}$$

Substituting Eq. (6.73) into Eq. (6.71) and Eq. (6.74) into Eq. (6.72)

$$\nabla_t \times E_t + \nabla_t E_z \times i_z + \gamma i_z \times E$$
$$= \nabla_t \times E_t + i_z \times (\gamma E_t - \nabla_t E_z)$$
$$= -j\omega B \tag{6.75}$$

$$\nabla_t \times H_t + E_z \times i_z + \gamma i_z \times H$$
$$= \nabla_t \times H_t + i_z \times (\gamma H_t - \nabla_t H_z)$$
$$= j\omega\varepsilon E. \tag{6.76}$$

Here, the relationships between the transverse component and the waveguide axis direction component of the electromagnetic fields are given from Eqs. (6.75) and (6.76),

$$\nabla_t \times E_t = -j\omega\mu_z H_z i_z \tag{6.77}$$

$$\nabla_t \times H_t = j\omega\varepsilon E_z i_z. \tag{6.78}$$

Substituting Eqs. (6.70) and (6.77) into Eq. (6.71),

$$- j\omega\mu_z H_z i_z + i_z \times (\gamma E_t - \nabla_t E_z)$$
$$= -j\omega(\mu H_t + j\mu' i_z \times H_t + \mu_z H_z i_z).$$

Arranging each side of the expressions,

$$i_z \times \gamma E_t + j\omega\mu H_t - \omega\mu' i_z \times H_t = i_z \times \nabla_t E_z. \tag{6.79}$$

Further, by substituting Eq. (6.78) into Eq. (6.75),

$$j\omega\varepsilon E_z i_z + i_z \times (\gamma H_t - \nabla_t H_z)$$
$$= j\omega\varepsilon E_z i_z + i_z \times \gamma H_t - i_z \times \nabla_t H_z$$
$$= j\omega E.$$

Since the relation $j\omega E = j\omega\varepsilon(E_t + E_z i_z)$ holds, substituting it into the right-hand side of Equation (6.76), and arranging this relation, the following expression is derived:

$$i_z \times \nabla_t H_z = i_z \times \gamma H_t + j\omega\varepsilon E_z i_z - j\omega\varepsilon(E_t + E_z i_z)$$
$$= i_z \times \gamma H_t - j\omega\varepsilon E_t. \tag{6.80}$$

Here, after defining the values of k^2, k'^2, and K^2 in the following expressions, the transverse electric field E_t and the magnetic field H_t are considered so that they are represented only by the electric and magnetic fields of the z component, respectively:

$$k^2 = \omega^2 \varepsilon\mu, \quad k'^2 = \omega^2 \varepsilon\mu', \quad K^2 = \omega^2 \varepsilon\mu + \gamma^2. \tag{6.81}$$

As a result of some calculations starting from Eq. (6.80), first, the following expression about E_t is obtained:

$$(K^4 - k'^4)E_t = \nabla_t(K^2 \gamma E_z + \omega\gamma^2 \mu H_z)$$
$$+ j i_z \times \nabla_t \{\omega(K^2 \mu - k'^2 \mu')H_z - \gamma k'^2 E_z\}, \tag{6.82}$$

where the relationship $\gamma^2 \mu' = K^2 \mu' - k'^2 \mu$ is used (see Appendix 6.A.4).

Similarly, the following equation about H_t can be derived from Eq. (6.80):

$$(K^4 - k'^4)H_t$$
$$= \nabla_t(K^2\gamma H_z + k'^2\omega\varepsilon E_z) - ji_z \times \nabla_t(K^2\omega\varepsilon E_z + k'^2\gamma H_z)^2. \tag{6.83}$$

Then, from the divergence equations, $\nabla \cdot Ee^{\gamma z} = 0$ and $\nabla \cdot Be^{\gamma z} = 0$,

$$\nabla_t \cdot E_t = -\gamma E_z \tag{6.84}$$

$$\nabla_t \cdot B_t = -\gamma B_z. \tag{6.85}$$

Moreover,

$$\nabla_t \cdot H_t = -\gamma\frac{\mu_z}{\mu}H_z - \omega\varepsilon\frac{\mu'}{\mu}E_z. \tag{6.86}$$

First, by taking the divergence of both sides in Eqs. (6.82) and (6.83), respectively, and applying the relationships of Eqs. (6.84) and (6.86), the following expression can be derived:

$$\nabla_t^2 E_z + aE_z + bH_z = 0. \tag{6.87}$$

Here, $a = K^2 - \frac{k'^2\mu'}{\mu}$, $b = -\frac{\omega\gamma\mu'\mu_z}{\mu}$ (see Appendix 6.A.5).
In the same manner,

$$\nabla_t^2 H_z + cH_z + dE_z = 0. \tag{6.88}$$

Here, $c = \frac{K^2\mu_z}{\mu}$, $d = \frac{\omega\gamma\varepsilon\mu'}{\mu}$.
Equations (6.87) and (6.88) are basic equations for determining the electric and magnetic fields in the present case.

6.3.2 Derivation of Electromagnetic-Field Components

Next let us consider how to convert the wave equations for E_z and H_z in hybrid equations (6.87) and (6.88) into the independent wave equations separated by E_z and H_z, respectively. This can be done by expressing E_z and H_z using the new variables U_1 and U_2, as shown in Eqs. (6.89) and (6.90):

$$E_z = P_1U_1 + P_2U_1 \tag{6.89}$$

$$H_z = Q_1U_2 + Q_2U_2. \tag{6.90}$$

In order to keep the generality in Eqs. (6.89) and (6.90), the following condition should be imposed:

$$\begin{vmatrix} P_1 & P_2 \\ Q_1 & Q_2 \end{vmatrix} \neq 0. \tag{6.91}$$

By substituting (6.89) into (6.87), and also (6.90) into (6.88), and arranging these expressions, respectively,

$$P_1 \nabla_t^2 U_1 + (aP_1 + bQ_1)U_1 + P_2 \nabla_t^2 U_2 + (aP_2 + bQ_2)U_2 = 0 \tag{6.92}$$

$$Q_1 \nabla_t^2 U_1 + (cQ_1 + dP_1)U_1 + Q_2 \nabla_t^2 U_2 + (cQ_2 + dP_2)U_2 = 0. \tag{6.93}$$

In addition, the expressions are set as follows:

$$aP_1 + bQ_1 = S_1 P_1 \tag{6.94}$$

$$aP_2 + bQ_2 = S_2 P_2 \tag{6.95}$$

$$dP_1 + cQ_1 = S_1 Q_1 \tag{6.96}$$

$$dP_2 + cQ_2 = S_2 Q_2, \tag{6.97}$$

where S_1 and S_2 are the separation constants.

Using the relationships of Eqs. (6.94)–(6.97), Eqs. (6.92) and (6.93) are given by

$$P_1(\nabla_t^2 U_1 + S_1 U_1) + P_2(\nabla_t^2 U_2 + S_2 U_2) = 0 \tag{6.98}$$

$$Q_1(\nabla_t^2 U_1 + S_1 U_1) + Q_2(\nabla_t^2 U_2 + S_2 U_2) = 0. \tag{6.99}$$

Also, if the relationship of Eq. (6.91) satisfies Eq. (6.98) and (6.99), the following expressions are obtained:

$$\nabla_t^2 U_1 + S_1 U_1 = 0 \tag{6.100}$$

$$\nabla_t^2 U_2 + S_2 U_2 = 0. \tag{6.101}$$

From Eq. (6.94),

$$(a - S_1)P_1 + bQ_1 = 0. \tag{6.102}$$

From Eq. (6.96),

$$dP_1 + (c - S_1)Q_1 = 0. \tag{6.103}$$

The condition which does not have trivial solutions in these simultaneous equations,

$$\begin{vmatrix} a - S_1 & b \\ d & c - S_1 \end{vmatrix} = 0. \tag{6.104}$$

Similarly, from Eq. (6.95) and (6.97),

$$\begin{vmatrix} a - S_2 & b \\ d & c - S_2 \end{vmatrix} = 0. \tag{6.105}$$

Since these equations have a difference only in S_1 and s_2, it becomes possible to satisfy Eqs. (6.104) and (6.105) by obtaining the solution of the quadratic equation based on Eq. (6.106):

$$\begin{vmatrix} a-S & b \\ d & c-S \end{vmatrix} = 0. \tag{6.106}$$

Hence, S_1 and s_2 are determined from this equation (6.106).

For convenience, by putting $P_1 = S_1$ and $P_2 = S_2$, the following expressions can be obtained from Eqs. (6.94) and (6.95):

$$Q_1 = \frac{(S_1 - a)S_1}{b}, \quad Q_2 = \frac{(S_2 - a)S_2}{b}.$$

Using these relationships, let us denote the coefficients in Eqs. (6.82) and (6.83), using S_1, S_2, U_2, and U_1.

As a result,

$$K^4 - k'^4 = \frac{\mu}{\mu_z} S_1 S_2, \tag{6.107}$$

(See Appendix 6.A.6.)

$$K^2 \gamma E_z + \omega \gamma^2 \mu' H_z = \gamma \frac{\mu}{\mu_z} S_1 S_2 (U_1 + U_2) \tag{6.108}$$

$$\omega(K^2 \mu - k'^2 \mu') H_z - \gamma k'^2 E_z = \omega \mu S_1 S_2 \left\{ \frac{(S_1 - a)U_1}{b} + \frac{(S_2 - a)U_2}{b} \right\} \tag{6.109}$$

$$K^2 \gamma H_z + k'^2 \omega \varepsilon E_z = \gamma S_1 S_2 \left\{ \frac{(S_1 - K^2)U_1}{b} + \frac{(S_2 - K^2)U_2}{b} \right\} \tag{6.110}$$

$$K^2 \omega \varepsilon E_z + k'^2 \gamma H_z = \omega \varepsilon \frac{\mu}{\mu_z} S_1 S_2 (U_1 + U_2). \tag{6.111}$$

Substituting Eqs. (6.107)–(6.109) into Eq. (6.82),

$$\frac{\mu}{\mu_z} S_1 S_2 E_t = \nabla_t \gamma \frac{\mu}{\mu_z} S_1 S_2 (U_1 + U_2)$$

$$+ j i_z \times \nabla_t (\omega \mu S_1 S_2) \frac{(S_1 - a)U_1}{b} + \frac{(S_2 - a)U_2}{b}.$$

Dividing both sides of this equation in $(\mu/\mu_z)S_1 S_2$,

$$E_t = \gamma \nabla_t (U_1 + U_2) - \frac{j\mu}{\gamma \mu'} i_z \times \nabla_t \{(S_1 - a)U_1 + (S_2 - a)U_2\}. \tag{6.112}$$

Substituting Eqs. (6.107), (6.110), and (6.111) into the Eq. (6.83),

$$\frac{\mu}{\mu_z} S_1 S_2 H_t = \nabla_t \gamma S_1 S_2 \left\{ \frac{(S_1 - K^2)U_1}{b} + \frac{(S_2 - K^2)U_2}{b} \right\}$$

$$- j i_z \times \nabla_t \omega \varepsilon \frac{\mu}{\mu_z} S_1 S_2 (U_1 + U_2).$$

Dividing both sides of this equation by $(\mu/\mu_z)S_1 S_2$,

$$H_t = \frac{1}{\omega\mu'}\nabla_t\{(K^2 - S_1)U_1 + (K^2 - S_2)U_2\} - j\omega\varepsilon i_z \times \nabla_t(U_1 + U_2). \quad (6.113)$$

Consequently, as shown by Eqs. (6.89), (6.90), (6.112), and (6.113), the electromagnetic fields in the waveguide axis direction and the transverse direction when mounting ferrite in a circular waveguide and applying H_{DC}, can be summarized as follows:

$$E_z = S_1 U_1 + S_2 U_2 \quad (6.114a)$$

$$H_z = \frac{S_1 - a}{b}S_1 U_1 + \frac{S_2 - a}{b}S_2 U_2 \quad (6.114b)$$

$$E_t = \gamma\nabla_t(U_1 + U_2) - \frac{j\mu}{\gamma\mu'}i_z \times \nabla_t\{(S_1 - a)U_1 + (S_2 - a)U_2\} \quad (6.114c)$$

$$H_t = \frac{1}{\omega\mu'}\nabla_t\{(K^2 - S_1)U_1 + (K^2 - S_2)U_2\} - j\omega\varepsilon i_z \times \nabla_t(U_1 + U_2). \quad (6.114d)$$

When the waveguide radius is ρ_0, the boundary conditions at the circular waveguide surface must be determined so that the tangential component of the electric field becomes 0. That is,

$$E_z = 0, \quad E_\varphi|_{\rho=\rho_0} = 0.$$

The second of the above-given conditions is obtained by substituting the relationship in Eq. (6.107) and the relationship $b = -\omega\gamma\mu_z\mu'/\mu$ in Eq. (6.87) into the Eq. (6.112).

As a result, the boundary conditions of the present circular waveguide are derived as follows:

$$S_1 U_1 + S_2 U_2 = 0 \quad (E_z|_{\rho=\rho_0} = 0). \quad (6.115)$$

$$\frac{\gamma}{\rho_0}\frac{\partial}{\partial\varphi}(U_1 + U_2) - \frac{j\mu}{\gamma\mu'}\frac{\partial}{\partial\varphi}\{(S_1 - a)U_1 + (S_2 - a)U_2\}|_{\rho=\rho_0} = 0. \quad (6.116)$$

6.3.3 Circular Waveguide with Ferrite

6.3.3.1 Ferrite Fully Filled Case

In this section, the electromagnetic-field analysis example is described in detail when filling with ferrite in a circular waveguide by applying the expressions derived in the previous section. When ferrite is filled into a circular waveguide, and a DC magnetic field H_{DC} is applied in the waveguide z-axis direction, as shown in the previous Figure 6.12, the characteristic equation for determining the propagation constant can be derived as follows.

Representing the radius ρ in the circular waveguide, and also the φ-direction dependence of the field by $e^{jn\varphi}$, the solutions of U_1 and U_2 which satisfy the wave equations (6.100) and (6.101) in the previous section are given by the following equations [9]:

$$U_1 = A_1 J_n(S_1^{1/2}\rho)e^{jn\varphi} \tag{6.117}$$

$$U_2 = A_2 J_n(S_1^{1/2}\rho)e^{jn\varphi}. \tag{6.118}$$

Here, $J_n(x)$ denotes the Bessel function of order n, and A_1 and A_2 are constants [9]. Applying boundary conditions (6.115) and (6.116),

$$S_1 J_n(S_1^{1/2}\rho_0)A_1 + S_2 J_n(S_2^{1/2}\rho_0)A_2 = 0 \tag{6.119}$$

$$\frac{jn\gamma}{\rho_0}\{J_n(S_1^{1/2}\rho_0)A_1 + J_n(S_2^{1/2}\rho_0)A_2\}$$

$$-\frac{j\mu}{\gamma\mu'}(S_1 - a)S_1^{1/2}J_n'(S_1^{1/2}\rho_0)A_1 + (S_2 - a)S_2^{1/2}J_n'(S_2^{1/2}\rho_0)A_2 = 0. \tag{6.120}$$

In order to have a nontrivial solution in Eqs. (6.119) and (6.120), the coefficient matrix equation for A_1 and A_2 must be zero.

In view of this, the following characteristic equation can be derived:

$$\mu\left[\left(1 - \frac{a}{S_1}\right)(S_1\rho_0)^{1/2}\frac{J_n'\{(S_1\rho_0)\}^{1/2}}{J_n\{(S_1\rho_0)\}^{1/2}} - \left(1 - \frac{a}{S_2}\right)(S_2\rho_0)^{1/2}\frac{J_n'(S_2\rho_0)^{1/2}}{J_n(S_2\rho_0)^{1/2}}\right]$$

$$+ n\mu'\gamma^2\left(\frac{1}{S_1} - \frac{1}{S_2}\right) = 0. \tag{6.121}$$

Here, S_1 and S_2 are given from Eq. (6.122) as the roots of the quadratic equation:

$$S_i^2 - (a + b)S_i + ac - bd = 0. \tag{6.122}$$

Since the values a, b, c, and d are defined in the previous expressions related to Eqs. (6.87) and (6.88), the propagation constant γ can be given by solving for S_i.

The Newton–Raphson method is one of the effective candidates for solving such a characteristic equation on a computer.

6.3.3.2 Ferrite Partially Filled Case

The characteristic equation that ferrite is partially mounted in a circular waveguide can be derived in the same manner as in the previous section. The present configuration is shown in Figure 6.13. The region of an isotropic dielectric medium $\rho_1 \leq \rho \leq \rho_0$ is represented using the subscript 0, and the ferrite region $0 \leq \rho \leq \rho_1$ is represented using the subscript 1.

In the dielectric medium ($\rho_1 \leq \rho \leq \rho_0$), the electromagnetic fields which are necessary for the boundary conditions are given by the following equations:

$$E_{z1} = A_2\left\{\frac{J_n(K_1\rho)}{J_n(K_1\rho_0)} - \frac{N_n(K_1\rho)}{N_n(K_1\rho_0)}\right\}e^{jn\varphi} = A_3 F(\gamma)e^{jn\varphi} \tag{6.123}$$

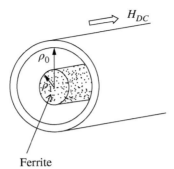

Figure 6.13 The case of loading a ferrite in a part of a circular waveguide.

$$H_{z1} = A_4 \left\{ \frac{J_n(K_1\rho)}{J_n'(K_1\rho_0)} - \frac{N_n(K_1\rho)}{N_n'(K_1\rho_0)} \right\} e^{jn\varphi} = A_4 G(\gamma) e^{jn\varphi}. \tag{6.124}$$

Here, $N_n(x)$ is the nth order Bessel function of the second kind and $K_1^2 = \omega^2 \varepsilon_1 \mu_1 + \gamma^2$. The prime (') means differentiation.

Next, in the ferrite medium ($0 \leq \rho \leq \rho_1$),

$$E_z = \{S_1 J_n(S_1^{1/2}\rho)A_1 + S_2 J_n(S_2^{1/2}\rho)A_2\} e^{jn\varphi} \tag{6.125}$$

$$H_z = \left\{ \frac{(S_1 - a)}{b} S_1 J_n(S_1^{1/2}\rho)A_1 + \frac{(S_2 - a)S_2}{b} J_n(S_2^{1/2}\rho)A_2 \right\} e^{jn\varphi}. \tag{6.126}$$

By applying the boundary condition that the tangential components of the electric and magnetic fields are equal at $\rho = \rho_1$, the four homogeneous equations can be obtained. From these homogeneous equations, the following equation is derived by imposing the condition that the coefficient determinant for the coefficients A_1, A_2, A_3, and A_4 must be 0.

That is,

$$P_1 Q_2 - P_2 Q_1 = 0. \tag{6.127}$$

Here, as for the subscripts of 1 and 2 in Eq.(6.127), the following equations are applied:

$$P_i = \{n\gamma^2 \mu' \mu_z (K_1^2 - S_i)G(\rho_1) + \mu_1 \mu \rho_1 G'(\rho_1)S_i(S_i - a)\}J_n(S_i^{1/2}\rho_1) \\ - \mu \mu_z K_1^2 G(\rho_1)(S_i - a)S_i^{1/2}\rho_1 J_n'(S_i^{1/2}\rho_1) \tag{6.128}$$

$$Q_i = \{n(K^2 - S_i)K_1^2 F(\rho_1)\mu_z \\ + n\mu S_i(S_i - a)F(\rho_1) + \omega^2 \varepsilon_1 \mu' \mu_z \rho_1 F'(\rho_1)S_i\}J_n(S_i^{1/2}\rho_1). \tag{6.129}$$

If the material constants in each medium are given, the propagation constant can be determined by numerical analysis, using Eq. (6.127).

6.3.4   Coaxial Waveguide with Ferrite

With reference to the method of circular waveguide analysis, let us introduce here the analytical result of propagation characteristics, when ferrite is loaded into a coaxial waveguide [10, 11]. We will investigate a ferrite medium with an infinite length to which an external DC magnetic field is applied in the coaxial waveguide axis direction, which coincides with the direction of EM-wave propagation, as shown in Figure 6.14.

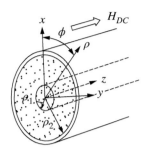

Figure 6.14 The case of loading a ferrite fully in a coaxial waveguide.

H_{DC} represents the total DC magnetic field in a ferrite medium. As a DC magnetic field value H_{DC} is to be applied, the strength of H_{DC} is assumed to range from the unsaturated region to the almost saturated region in the present case. In addition, a quasi-TEM wave analysis is assumed.

The reason for investigating here the coaxial waveguide is that it is extremely difficult to measure EM-wave attenuation characteristics and absorption characteristics when applying a DC magnetic field to a flat ferrite in free space in many cases. In other words, this is because the investigation of attenuation characteristics and EM-wave absorption characteristics at the time of an incident plane wave (quasi-TEM wave) can be simplified by the method of using a coaxial waveguide.

According to the theory described so far, we can derive a characteristic equation for obtaining a propagation constant in the case where the ferrite is fully or partially mounted in the coaxial waveguide. The propagation constants can be obtained by computer analysis of the characteristic equation based on the Newton–Raphson method [10]. In the following, the examples of analysis for propagation constants when ferrite is filled in a coaxial waveguide are taken up, and the main points of this kind of analysis and the significance of this type of analysis are described. Now, Figure 6.15 denotes the analysis result of the attenuation constant [11].

In order to validate the theoretical analysis in Figure 6.15, the present analytical results are compared with the analytical results calculated using the expressions of Suhl and Walker, which were used to approximate evaluation of this kind of the propagation characteristics [12]. An approximate expression of Suhl and Walker is derived from the assumption that the distance between the upper and lower conductive plates in the slab waveguide is sufficiently narrow compared to the wavelength. In this slab waveguide case, a static magnetic field, of course, is applied to the ferrite in the EM-wave propagation direction. Suhl and Walker's approximate expression of propagation constant is given using

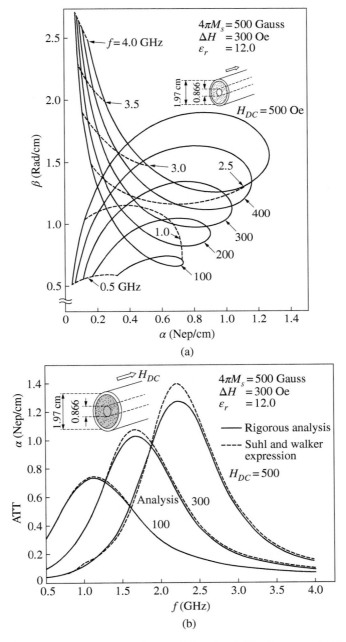

Figure 6.15 (a) Propagation constant characteristics when taking frequency (dotted line) and applied DC magnetic field as parameters. (b) Comparison of the rigorous analytical result for attenuation characteristics in a ferrite-filled coaxial waveguide, and the approximation formula of Suhl and Walker when H_{DC} is taken as a parameter. (c) Attenuation characteristics when H_{DC} and $4\pi M_s$ are taken as parameters.

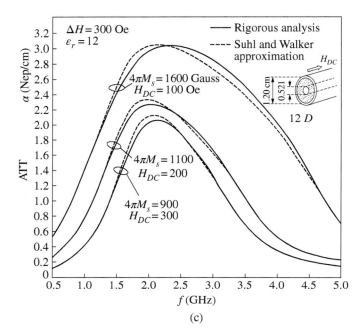

Figure 6.15 (*Continued*)

the effective magnetic permeability. An approximate expression for the propagation constant by Suhl and Walker is represented as follows:

$$\Gamma = j\frac{\lambda}{2\pi}\sqrt{\varepsilon_r\mu_{eff}},\tag{6.130}$$

where $\mu_{eff} = \frac{\mu^2 - \mu'^2}{\mu^2}$, and μ, μ' are the tensor magnetic permeability.

Figure 6.15a presents propagation characteristics when taking H_{DC} as a parameter, while keeping saturation magnetization $4\pi M_S$ and the magnetic resonance half-width ΔH as the constants. Figure 6.15b shows the frequency characteristics of attenuation when taking H_{DC} as a parameter, keeping $4\pi M_S$ and ΔH constant. As shown in Figure 6.15b, as the applied static magnetic field H_{DC} is increased, it can be seen that the errors occur at the attenuation constant values between those of the Suhl–Walker approximation analysis and the strict cylindrical coordinate analysis.

In Figure 6.15b,c, the dotted line is the analytical result of the attenuation constant by Suhl and Walker's approximate expression in Eq. (6.130), and the solid line represents the exact theoretical analysis result due to the present theory. Figure 6.15a represents the propagation characteristics when taking H_{DC} as a parameter, while keeping saturation magnetization $4\pi M_S$ and magnetic resonance half-width ΔH as constant values. Figure 6.15b shows the frequency characteristics of attenuation when taking H_{DC} as a parameter, keeping $4\pi M_S$ and ΔH as constant values. As shown in Figure 6.15b, as the applied static magnetic field H_{DC} is increased, it can be seen that errors occur in the attenuation constant values between the analysis by the Suhl–Walker approximation and the strict cylindrical coordinate analysis.

From these analyses, it is found that the attenuation constant α increases when increasing the value of H_{DC} under the conditions of keeping $4\pi M_S$ and ΔH constant, and it shifts to the region of higher frequencies. Figure 6.15c investigates the mutual relations between the values of H_{DC} and $4\pi M_S$ against the attenuation constant α, while keeping the value of ΔH constant. We should note here that the propagation characteristics of a magnetized ferrite medium can often be evaluated using the values of $4\pi M_S$ and ΔH without using each tensor permeability element. In other words, even if the individual tensor values are not investigated, it becomes possible to understand the present attenuation behaviors intuitively from the viewpoint of $4\pi M_s$ and ΔH values.

Also, it should be noted here that these models of analysis are set in cases where the ferrites are mounted in the infinite direction of the waveguides. However, when conducting the experiment to measure the attenuation constant, the ferrite size in the direction of the coaxial waveguide axis takes a finite value. Therefore, the net DC magnetic field strength applied to the finite ferrite cylinder length is different from the actually applied external magnetic DC field under the influence of a demagnetizing field. To avoid this problem, it is necessary to take into account the minimization of the influence of the demagnetizing field by taking the length of the ferrite cylinder in the wave propagation direction as long as possible.

Further, when comparing the theoretical attenuation constant value with its measured value, we have to consider a matching problem of the ferrite loaded in the coaxial waveguide. In this regard, a configuration in which both ends of a ferrite cylinder are cut obliquely is effective. Then, the effect of magnetic resonance half-width ΔH is presumed from its resonance nature beforehand, that is, it can be found that it contributes to the magnitude of the frequency band of the attenuation curve. The above is an example of theoretical analysis intended to be applied to a coaxial attenuator, or the like.

As for the EM-wave absorber, when a static magnetic field is applied perpendicular to the ferrite surface and at the same time the ferrite thickness varies depending on the value of H_{DC}, a thinner EM-wave absorber can be realized, as described in Chapter 5 [13]. As a simple measurement method in such a case, there exists a method of attaching a flat donut-shaped ferrite to the short-circuited terminal end of a coaxial waveguide and measuring the electromagnetic-wave absorption characteristic by applying H_{DC} in the waveguide axial direction [13]. In this section, the purpose of emphasizing the method of using a coaxial waveguide is to avoid the difficulty of space measurement method in this kind. That is, the coaxial waveguide method becomes effective when measuring the anisotropic EM-wave absorber during application of H_{DC}.

By the way, it is not easy to specify the material constants that show the best characteristic, using only the measurement methods in the actual production of EM-wave absorbers and microwave attenuators when H_{DC} is applied. When plane-shaped EM-wave absorbers and coaxial attenuators are designed, the propagation constant analysis using the cylindrical coordinate system enables various parameter analyses and it consequently helps simplify the measurement.

6.4 Metamaterial

Since the late 1900s, the materials that simultaneously have negative permittivity and negative permeability began to attract attention. This is called metamaterial because this kind of material cannot exist in nature. Usually, this is called electromagnetic metamaterials. Because this material has the abovementioned specificity, it is expected that it will allow the creation of a new technical field that will overcome the conventional material concepts and characteristics of microwave devices. In this section, we mainly summarize the basic theories of artificial materials, which simultaneously have negative dielectric constant and magnetic permeability, as well as examples of applications for EM-wave absorbers

6.4.1 Metamaterial Outlines

Generally, metamaterials are classified into two types. One of them is a material which simultaneously takes the negative permittivity and permeability described earlier, and also takes a negative refractive index against electromagnetic waves. The metamaterial of this kind is called a "left-handed metamaterial." Another type is a material composed of artificial substance in a broader sense, and this is called a "right-handed metamaterial." The fundamental idea of realization of these types of metamaterials is to realize electric and magnetic dipoles in the medium and make them resonate using electromagnetic waves.

In general, the history of metamaterials is old and dates back to the 1940s [14, 15].

In the mid-1940s, an artificial dielectric material composed of periodic metallic structures was already proposed. The examples of the proposed metamaterials of this type are the minute metallic balls, strip-like metal, thin metal wires, and split ring, as shown in Figure 6.16 [14]. As an example, this kind of material was used instead of a conventional dielectric material lens by mounting the strip-like periodical metal arrangement as a lens in front of the

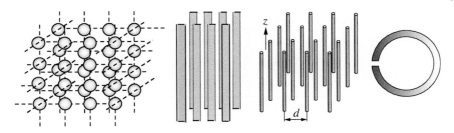

Figure 6.16 Artificial material in the early days.

parabolic antenna. That is, compared with the dielectric material used for the lens of a microwave parabolic antenna, conventionally, this type of artificial material was adopted due to the excellent refractive index condition required and satisfying the weight reduction condition.

In 1967, from the theoretical stand point, the Russian physicist V. G. Veselago presented a thesis on the existence of a medium whose permittivity and permeability become negative simultaneously [16].

Later in the 1990s, J. B. Pendry et al. proposed a specific structure consisting of the metal which can realize negative permittivity and negative permeability [17, 18]. At this time, thin metal wires and double split rings, shown in Figure 6.17, were in the objective of the analysis. As a response to this, the research was gaining attention to the artificial materials as "a material that cannot exist in nature," and the metamaterials research has activated again. During 2000–2001, the existence of the left-handed medium was verified experimentally by D. R. Smith, R. A. Shelby et al. [19, 20].

In this experiment, periodic structures consisting of thin metal wires and double split-ring elements, which were proposed by Pendry, were used. This fact again triggered attention to artificial materials, and the metamaterials research became active, particularly in the left-handed metamaterials. In 2002, A. Sanada, C. Caloz and T. Itoh, G. V. Eleftheriades, and A. A. Oliner

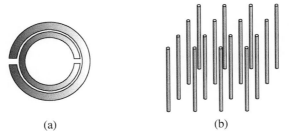

(a)   (b)

Figure 6.17 Metal model introduced by Pendry in the medium composition. (a) Double-split ring. (b) Thin metal wires.

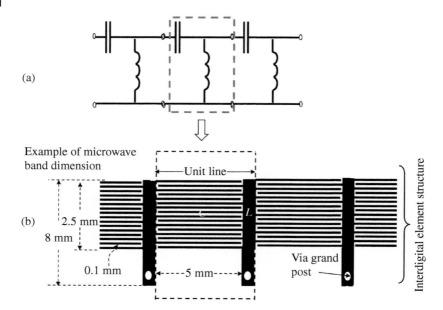

Figure 6.18 Interdigital circuit.

et al., published papers where a "left-handed metamaterial" is transformed into "circuit concept" at almost the same time [21–26]. This equivalent circuit is implemented on the transmission line using the capacitor and inductor circuit elements, and these combinations are called the interdigital structure, as shown in Figure 6.18 [24]. The opportunity to shift to this circuit concept led to the emergence of significant new technologies, especially in the fields of microwave circuit devices and antenna applications, and the like, in the left-handed metamaterial researches.

Generally, the left-handed metamaterial can be defined as a material that can behave as a uniform medium equivalently for electromagnetic waves, and a material that consists of a structure with electrical and magnetic properties, which do not exist in nature.

Here, when defining an equivalent medium which is equal to a "homogeneous medium," this condition is usually limited to the case when the relationship of $P < \lambda/4$ is held. Here, P is the dimension of the unit cell that is configured. Meanwhile, the right-handed metamaterial is defined as a material exhibiting functional properties that is configured on the basis of an engineering method and material that is configured through nanotechnology and chemical materials with properties that cannot exist inherently. Therefore, the right-handed metamaterial includes artificial material in a so-called broader sense. For example, the carbon nanotube material can be considered as belonging to the right-handed material.

6.4.2 Metamaterial Theories

6.4.2.1 Left-Handed and Right-Handed Systems

The origins of the etymology of the left-handedness and right-handedness are based on the vector definition. When the electric field, the magnetic-field vector and the wave vector are represented by E, H, k, respectively, as shown in Figure 6.19, if this relation is "counterclockwise," this material is defined as the right-handed system. And, when this relation is "clockwise," this material is defined as the left-handed system.

First of all, let us clarify theoretically the tripartite relations of E, H, and k, which depend on the value of the dielectric constant and magnetic permeability in a lossless medium.

The electromagnetic field of a plane wave propagating in the x-axis direction, shown in Figure 6.20, is given by the following equation:

$$E = E_{0y}e^{j(\omega t - kx)} = i_y E_{0y}e^{j(\omega t - kx)} \tag{6.131}$$

$$H = \frac{E}{\eta} = \frac{E_0 e^{j(\omega t - kx)}}{\eta} = \frac{1}{\eta}i_y E_{0y}e^{j(\omega t - kx)}. \tag{6.132}$$

The Maxwell's equations are

$$\nabla \times E = -\frac{\partial B}{\partial t} \tag{6.133}$$

$$\nabla \times H = \frac{\partial D}{\partial t}. \tag{6.134}$$

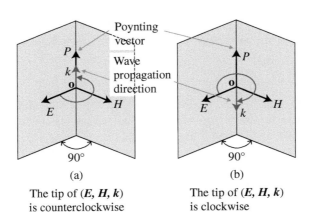

(a) The tip of (E, H, k) is counterclockwise

(b) The tip of (E, H, k) is clockwise

Figure 6.19 (a) Relation of (E, H, k) in right-handed system. (b) Relation of (E, H, k) in light-handed system. (E, H, K) is the triad of Electric field, magnetic field, and wave vector, respectively. P is Poynting vector

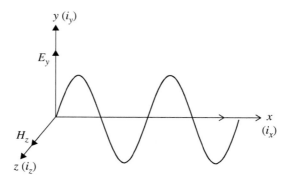

Figure 6.20 Electric field distribution.

In this case, electric fields are assumed to have the relationships $E_x = E_z = 0$, $\partial/\partial z = 0$, so that, substituting Eq. (6.131) into (6.133), we get

$$\nabla \times E = \left(i_x \frac{\partial}{\partial x} + i_y \frac{\partial}{\partial y} \right) \times i_y E_{0y} e^{-jkx} = -j(i_x k) \times i_y E_{0y} e^{-jkx}$$

$$= -jk \times E = -j\omega\mu H \quad \therefore k \times E = +\omega\mu H. \tag{6.135}$$

Similarly, from Eq. (6.132) and (6.134),

$$\nabla \times H = \left(i_x \frac{\partial}{\partial x} + i_y \frac{\partial}{\partial y} \right) \times \frac{1}{\eta} i_y E_{0y} e^{j(\omega t - kx)} = -\frac{j}{\eta}(i_x k) \times i_y E_{0y} e^{j(\omega t - kx)}$$

$$= -\frac{j}{\eta}(k \times E) = -\frac{j}{\eta}(k \times \eta H) = j\omega\varepsilon E \quad \therefore k \times H = -\omega\varepsilon E. \tag{6.136}$$

Now, when the relations $\varepsilon, \mu > 0$ are held, that is, when the permittivity and permeability become positive simultaneously, the following expressions yield:

$$k \times E = +\omega\mu H$$
$$k \times H = -\omega\varepsilon E. \tag{6.137}$$

This means the right-handed relationship in each vector E, H, and k, as shown in Figure 6.19a. On the other hand, when the relations $\varepsilon < 0$, $\mu < 0$ are held, that is, when the dielectric constant and magnetic permeability are simultaneously negative, the following expressions are provided. In this case, it is confirmed that the left-handed relationship in Figure 6.19b is satisfied [24].

$$\mu = -|\mu| \quad k \times E = -\omega|\mu|H$$
$$\varepsilon = -|\varepsilon| \quad k \times H = +\omega|\varepsilon|E. \tag{6.138}$$

6.4.2.2 Conversion from Material to Transmission Line Concept

In this section, let us briefly describe the fact that the relationship between the wave number vector k and the Poynting vector P turns out to be antiparallel.

In the LH materials shown in Figure 6.19b, the wave number vector **k** in the direction of EM-wave propagation and the Poynting vector **P** take mutually opposite directions.

This phenomenon, at the depiction method in Figure 6.21, means that the vector orientations of the "phase velocity V_p" and "group velocity V_g" are in the opposite direction from each other in the fourth quadrant.

This theoretical explanation becomes easily understandable from viewpoint of the theory of circuits. Next, let us consider how to transform the left-handed

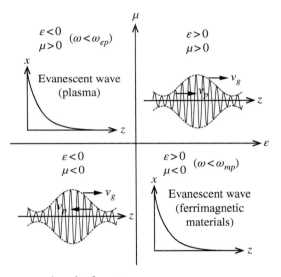

ω : Angular frequency
ω_{ep}, ω_{mp} : Electrical, magnetic plasma angular frequency
v_g : Group velocity
v_p : Phase velocity
ε : Dielectric constant
μ : Magnetic permeability

Figure 6.21 Dielectric values, permeability values, and propagation forms.
(a) The relative permittivity ε_r in the case where charged particles do not collide with other neutral atoms etc. in the plasma medium has the following relationship with the plasma angular frequency.

$$\varepsilon_r = 1 - \omega_{ep}^2 / \omega^2$$

where, ω_{ep} is plasma angular frequency, ω is EM–wave angular frequency.
We can understand from this equation that the dielectric constant is negative when the plasma angular frequency ω_{ep} is higher than the angular frequency in EM -waves.
(b) The Ferrimagnetic material described in the fourth quadrant means the ferrite described in Chapter 5. Ferrite has a negative permeability when a static magnetic field is applied or it has anisotropy.

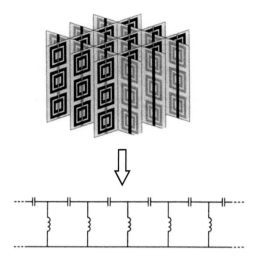

Figure 6.22 Conversion of the material medium concept to an equivalent transmission line concept.

metamaterial concept into the concept of equivalent transmission line circuit, as shown in Figure 6.23.

By the way, the electromagnetic-wave propagation that the phase velocity and the group velocity become the antiparallel is not new. In the late 1960s, L. Brillouin, who was concerned with the wave propagation of the periodic structure transmission line, and J. R. Pierce, who was involved in the transmission line problem on the traveling-wave tube, had already found the phenomena of causing antiparallel waves. Therefore, Figure 6.22 has already been presented as a circuit that can produce the backward wave.

This circuit consists of the same circuit element configuration as the pure left-handed transmission line, which is discussed in the present metamaterial. However, in practical left-handed metamaterial circuits, this concept is extended further, as shown in Figure 6.23. This is because, in practical use, it is necessary to take into account the L_R and C_R circuit elements due to parasitic on the conventional right-handed high-frequency line. Therefore, when dealing with the transmission lines representing metamaterials, this composite right-handed/left-handed transmission line (lossless) (CRLHTL) should be considered, as shown in Figure 6.23a [24].

In the case when the loss of TL is taken into consideration, the equivalent TL line of the left-handed metamaterial can be shown in Figure 6.23b [24].

6.4.3 Negative Permittivity and Permeability

Conventionally, the ferrite material and carbon materials have been used as EM-wave absorber materials. In contrast, left-handed metamaterials have a

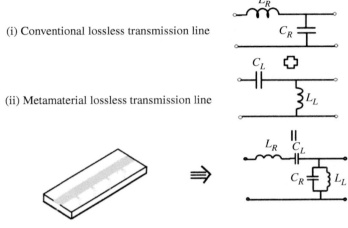

(i) Conventional lossless transmission line

(ii) Metamaterial lossless transmission line

(iii) Metamaterial right / left handed lossless transmission line

(a)

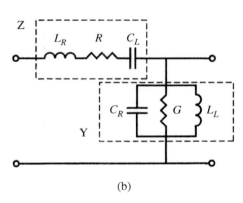

(b)

Figure 6.23 Right-/left-handed composite equivalent transmission line. (Subscripts R and L mean right-handed and left-handed, respectively.). (a) Composite right-/left-handed lossless equivalent transmission line. (b) Composite right-/left-handed transmission line with loss.

great feature in that effective permittivity and permeability are realized, using only common metals (inexpensive general-purpose metals such as copper). By the way, effective permittivity and permeability are due to the polarization dependence on the shape of each metal element. Namely, to say it simply, they arise from the phenomenon of electric and magnetic polarization. If we want to easily grasp this phenomenon based on a quasi-electrostatic field, this is considered as follows.

When the electric field E of the incident wave is concentrated to a sufficiently small spherical metal element in comparison with the wavelength, an electric dipole arises. And, also, when the magnetic field H of the incident wave concentrates on a sufficiently small spherical metallic element, the magnetic dipole arises from an eddy current. Thus, it is considered that the nature of the permittivity and permeability essentially depends on the behavior of the dipoles in the material.

In the example of a single-split ring in Figure 6.16 on the right side, if the split ring is sufficiently small compared to the wavelength of the incident wave electric field E, an electric polarization arises. Also, if the ring size is adjusted so that the circulating current I_c is induced mainly by the incident magnetic field H, magnetic polarization arises. Thus, the nature of permittivity and permeability is based, essentially, on the resonance characteristics of electric and magnetic dipoles in the material under consideration.

Now let us introduce the equations of effective permittivity and permeability, which were derived by J. B. Pendry et al., and their significance [17, 18]. J. B. Pendry et al. derived expressions for the effective permittivity from the concept of plasmon in the field of electronic physics.

Plasmon is defined as follows:

First, vibration due to a fluctuation in the electron density is called plasma vibration.

When regarding the wave propagation by this plasma vibration as a pseudo particle with momentum of hk and energy of $h\omega_p$, this state is called plasmon. The following equation is the effective permittivity derived by Pendry et al, in the case where the conductive thin wire model is used, as shown in Figure 6.24;

$$\varepsilon_r(\omega) = 1 - \frac{\omega_p^2}{\omega(\omega + i\gamma)} = 1 - \frac{\omega_p^2(\omega - i\gamma)}{\omega(\omega + i\gamma)(\omega - i\gamma)}$$

$$= 1 - \frac{\omega_p^2}{\omega^2 + \gamma^2} + j\frac{\gamma\,\omega_p^2}{\omega(\omega^2 + \gamma^2)}. \tag{6.139}$$

Here, ω is the angular frequency, ω_p is the electric plasma angular frequency $\omega_p^2 = 2\pi c_0/a^2 \ln(a/r)$ (c_0 is the speed of light, r is the radius of wires, a is the distance between two wires), γ is the damping factor, and $\gamma = \varepsilon_0(a\omega_p/r)^2/\pi\sigma$ (σ is the conductivity) .

Let us briefly explain the meaning of this expression. The following expression, which shows the negative dielectric constant, becomes easy to understand, considering separately the lossy and lossless media.

(i) *The case with the loss in a thin line metal* (see Figure 6.24):
From the first and the second terms on the right-hand side of Eq. (6.139), the real part of permittivity takes the negative value ($\mathrm{Re}(\varepsilon_r) < 0$) under condition $1 < \frac{\omega_p^2}{\omega^2+\gamma^2}$.
That is, $\omega^2 < \omega_p^2 - \gamma^2$

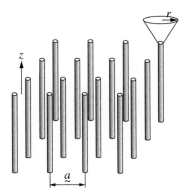

Figure 6.24 Conducting thin wires to explain the negative dielectric constant. σ: Conductivity, a: Thin wire spacing, r: Thin line radius.

(ii) *Lossless case in a thin line metal*:
From the second term on the right-hand side of Eq. (6.139), under the condition $\omega < \omega_p$, ε_r takes the negative value.

To summarize the derivation of Eq. (6.139),

(a) First, consider a unit cell model consisting of a thin metal wire (radius about 1 μm).
(b) The expression of the angular frequency in a normal plasma oscillation is given by

$$\omega_p^2 = \frac{ne^2}{\varepsilon_0 m}. \tag{6.140}$$

Here, we have to derive the relations represented by the following:

(a) The effective electron density n,
(b) The effective mass m in Eq. (6.140).
(c) As the next step, the following equation for the dielectric constant in the plasma medium should be calculated.

$$\varepsilon(\omega) = 1 - \frac{\omega_p^2}{\omega(\omega + i\gamma)}. \tag{6.141}$$

In this expression, a damping term γ, including the conductivity in Eq. (6.141), has to be derived. By substituting all these values (n, m, γ) into Eq. (6.141), we can derive Eq. (6.139). In the course of deriving Eq. (6.141), it is necessary to search for dipole moments per unit volume in the plasma medium.

On the other hand, the expression of the negative effective permeability follows from the idea that a magnetic dipole can be generated by a resonant current flowing in a double-split ring with a capacitive slot, as shown in Figure 6.25.

$$\mu_r(\omega) = 1 - \frac{F\omega}{\omega^2 - \omega_m^2 + i\omega\varsigma}$$

$$= 1 - \frac{F\omega^2(\omega^2 - \omega_m^2)}{(\omega^2 - \omega_m^2)^2 + (\omega\varsigma)^2}$$

$$+ i\frac{F\omega^2\varsigma}{(\omega^2 - \omega_m^2)^2 + (\omega\varsigma)^2}. \qquad (6.142)$$

Figure 6.25 Analysis model of double-split rings [24].

Here, $F = \pi(r/a)^2$, ω is the angular frequency, ω_m is the magnetic resonance frequency (effective magnetic plasma frequency), $\omega_m = c_0\sqrt{3a/\pi\ln(2cr^3/d)}$, ς is the damping factor, $\varsigma = (2l\sigma_1/r\mu_0)$, r is the inner radius of the smaller ring, a is the distance between the centers of two rings, c is the width of rings, c_0 is the speed of light, and d is the space between two rings in the radial direction.

In this effective magnetic permeability, even if the double-split metal is in the lossy or lossless case, of course, a magnetic plasma frequency exists which governs the negative effective permeability value. However, in many cases, when considering negative magnetic permeability, the lossless case is considered.

In the lossless case, when $\omega_m < \omega < \frac{\omega_m}{\sqrt{1-F}} = \omega_{pm}$, the effective permeability takes a negative value. ω_{pm} is the magnetic plasma frequency.

Although we have introduced here the effective dielectric constant and magnetic permeability based on the Pendry theory, we can, of course, derive these equations on the basis of Maxwell's equations [27, 28]. In this case, in order to obtain an effective permittivity and effective permeability with the Maxwell's equations as a theoretical basis, it is first of all necessary to pay attention to the cause of the electric polarization and magnetic polarization occurring in the metamaterial.

Then, by making the electric polarization and magnetic polarization phenomenon associated with the electric current source and magnetic current source, respectively, in Maxwell's equations, one can obtain the equation of effective permittivity and effective magnetic permeability incorporating the frequency response [27].

6.4.4 Negative Refractive Index Medium

As noted in the previous section, D. R. Smith and R. A. Shelby et al. experimentally proved the existence of a metamaterial using the structure, as shown in Figure 6.26. The sample used for the experiment is composed of a periodic structure of a combination of thin metal wires and double-split-ring elements, as shown in Figure 6.26. To realize negative magnetic permeability, double-split rings are created in a rectangular shape, and to achieve a negative dielectric constant, a configuration that is mounted with metallic thin wires on the back of the substrate is introduced. Using the 5 GHz band of an EM-wave incident on this medium, the existence of left-handed metamaterials was demonstrated by checking whether the refraction angle is negative or not [21].

Next, let us consider the negative refractive property from the theoretical viewpoint when the EM-wave is incident on the left-handed medium.

The phase constant is now expressed by the following equation:

$$\beta = \omega\sqrt{\varepsilon\mu} = \omega\sqrt{\varepsilon_0\mu_0}\sqrt{\varepsilon_r\mu_r}$$
$$= \frac{\omega\sqrt{\varepsilon_r\mu_r}}{c} = \frac{\omega n}{c}, \tag{6.143}$$

where c is the speed of light and n is the absolute refractive index.

The relative refractive index is the ratio of the refractive indices of two media. In particular, the refractive index with respect to vacuum (approximately the case of air) is called the absolute refractive index.

In the present case, it is assumed that medium I is the case of vacuum. Hence, the absolute refractive index of medium II is given in the following expression:

$$n = \frac{\sqrt{\varepsilon_2\mu_2}}{\sqrt{\varepsilon_1\mu_1}} = \frac{\sqrt{\varepsilon_2\mu_2}}{\sqrt{\varepsilon_0\mu_0}} = \sqrt{\varepsilon_{2r}\mu_{2r}}, \tag{6.144}$$

where,

$$\sqrt{\varepsilon_i\mu_i} = \sqrt{\varepsilon_0\mu_0}\sqrt{\varepsilon_{ir}\mu_{ir}} \quad i = 1, 2.$$

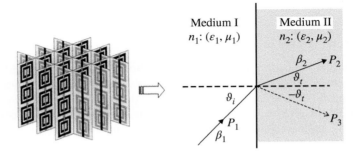

Figure 6.26 Negative refraction characteristic.

Moreover, the phase constant is expressed by the following equation:

$$\beta = \omega\sqrt{\varepsilon\mu} = \omega\sqrt{\varepsilon_0\mu_0}\sqrt{\varepsilon_r\mu_r}$$
$$= \frac{\omega\sqrt{\varepsilon_r\mu_r}}{C} = n\frac{\omega}{C}. \qquad (6.145)$$

That is,

$$\beta_1 = n_1\frac{\omega}{C}, \quad \beta_2 = n_2\frac{\omega}{C}. \qquad (6.146)$$

The following relationship holds in the left-handed medium:

$$n_1 > 0, \quad n_2 = -|n_2|$$

$$\frac{\sin\theta_i}{\sin\theta_t} = \frac{n_2}{n_1}.$$

Substituting this relation into the following Snell's law, from the relationship, in the left-handed medium,

$$\beta_2 < 0, \quad \therefore n_2 < 0(V_g > 0). \qquad (6.147)$$

Rewriting this relation,

$$n_1 > 0, \quad n_2 = -|n_2|. \qquad (6.148)$$

When substituting this expression into Snell's Law, the following relation is obtained:

$$|n_1|\sin\theta_i = -|n_2|\sin\theta_t = |n_2|\sin(-\theta_t). \qquad (6.149)$$

That is, the Snell's law in the left-handed medium can be given by

$$\frac{\sin\theta_i}{\sin(-\theta_t)} = \frac{n_2}{n_1}. \qquad (6.150)$$

Thus, it is found that the refraction angle takes a negative value in the left-handed medium.

6.4.5 Metamaterial as a Medium

In this section, as a basic knowledge of the method of constructing an EM-wave absorber, we shall outline about various kinds of metamaterial media. The left-handed metamaterials were made of conductive elements, as described in the previous section. However, on the other hand, attempts to construct left-handed metamaterials from conventional material concepts have also been investigated.

Since we can find that there exists an interesting point in this way of thinking, let us first consider this topic. The basis of the permittivity and permeability characteristics in an electrical material depends on the presence of electric and magnetic dipole moment, respectively, as described previously. An example of constructing a metamaterial using conventional materials was proposed by focusing on this dipole moment.

The idea underlying this kind of medium configuration was started with a theoretical study of the effective permittivity and effective permeability by Lewin [29]. Based on the theoretical idea of Lewin, Vendik attempted to realize a left-handed medium using both large and small spherical ferroelectrics with each radius of 100 and 70.5 μm and with a dielectric constant of 1000 [30].

These ferroelectrics are arranged at each vertex of a cubic lattice with a side length of 500 μm, and a resonance-type medium in a three-dimensional structure is realized. That is, this principle is based on the idea of creating two large and small dielectric spheres and generating dipole-like TM_{011} mode and TE_{011} mode within each dielectric sphere, as in the case of spherical cavities. This idea is interesting because it is not based on the idea of the principle of LH metamaterial implementations using metal units to realize an electric dipole and a magnetic dipole, but this LH metamaterial was achieved by using only conventional materials.

By the way, in the high-frequency range, such as the terahertz band, where technological progress is expected in the future, it is expected that it will be difficult to apply a magnetic material having sufficient magnetic characteristics necessary for various device designs. That is, conventional general-purpose ferrite is no longer available in this frequency band. As a magnetic material in the terahertz band, Yen et al. proposed a split-ring resonator (SRR) on a quartz substrate with a thickness of 600 μm using a photolithographic process [31]. The split ring is made of copper foil of 3 μm thickness. It is pointed out that the bandwidth of the reflectance ratio can be controlled by adjusting the SRR parameter. Also, it is reported that the resonance point of the reflectance ratio shifts from the low-frequency region to the high-frequency region as the size of the SRR increases. Further, it is pointed out that the bandwidth of the reflectance ratio can be controlled by adjusting the SRR parameter.

As another example of the LH metamaterial in the terahertz band, a periodic structure of rod-split-ring (RSR) consisting of Ni and Au has been proposed by Moser et al. [32]. In this case, the frequency band is 1–2.7 THz.

6.4.6 Metamaterial Absorber

Some EM absorbers, which are based on the idea of left-handed metamaterials, have also been proposed [33, 34]. An EM-wave absorber consisting of a metamaterial can be characterized at a point without a conductive plate on the back of the EM-wave absorber. For example, F. Bilotti et al. proposed an EM-wave absorber which uses a resistive film on the incident EM-wave surface and a rectangular split ring on its back [33]. This absorber is based on the principle that a resonant electric field of a rectangular split ring is absorbed by a resistive film in the front. In this case, of course, the reflective conductor plate is not loaded in the back of the absorber materials. The absorber thickness is only 5 mm at 2.05 GHz.

In contrast, Landy et al. proposed an EM-wave absorber by introducing a unit element structure which is called a paired nanorod structure, aiming at a full-wave absorber with 100% absorptivity [34]. The metamaterial unit element in this case is constructed by an electric-field coupling-type resonator in the form of a split ring (electric ring resonator, ERR) that shares a central conductor and a unit element called a cutting wire which are arranged on the back of the split ring.

In this configuration, it is reported that the magnetic response of the configuration element can be adjusted relatively easily by just adjusting the ERR shape and adjusting the space between the ERR and the cut wire. Therefore, it is considered that the effective dielectric constant and effective permeability can be regulated separately. The experimental absorbance characteristic value represents about 0.8 at a frequency of 11.5 GHz.

In this type of metamaterial wave absorber, since the conductor plate is not used on the back of the material, the wave absorption characteristics should be evaluated on the basis of the absorbance definition by measuring the transmitted wave together with the reflection characteristics. Here, it should be noted that the definition of absorbance is a concept that includes wave transmission, reflection, and scattering.

6.A Appendix

6.A.1 Appendix to Section 6.1.2 (1)

Magnetic field formula of chiral medium:

(a) *Incident magnetic field vector ($z < 0$) in medium I:*

$$H_i = H_{0i}e^{ik_0(z \cos \theta_0 - x \sin \theta_0)},$$

where

$$H_{0i} = \eta_0^{-1}[E_{i\|}e_y - E_{i\perp}(\cos \theta_0 e_x + \sin \theta_0 e_z)].$$

Reflected magnetic field vector ($z < 0$) in medium I:

$$H_r = H_{0r}e^{-ik_0(z \cos \theta_0 + x \sin \theta_0)},$$

where

$$H_{0r} = \eta_0^{-1}[-E_{r\|}e_y + E_{r\perp}(\cos \theta_0 e_x - \sin \theta_0 e_z)].$$

(b) *Magnetic field vector in chiral medium (medium II):*
    The magnetic fields in the chiral medium consist of the following fields:

(1) The sum H_c^+ of the clockwise circularly polarized wave and the counter-clockwise circularly polarized wave propagating to the boundary plane side $z = d$.

(2) The sum H_c^- of the clockwise circularly polarized wave and the counter-clockwise circularly polarized wave propagating to the boundary plane side $z = 0$.

In the same way of the electric field, the magnetic field is composed of these sums. That is,

$$H_c = H_c^+ + H_c^-.$$

At this time, H_c^+ and H_c^- are given by the following formulas:

$$H_c^+ = H_1^+ e^{ik_1(z\cos\theta_1 - x\sin\theta_1)}$$
$$+ H_2^+ e^{ik_2(z\cos\theta_2 - x\sin\theta_2)}$$
$$H_c^- = H_1^- e^{-ik_1(z\cos\theta_1 + x\sin\theta_1)}$$
$$+ H_2^- e^{-ik_2(z\cos\theta_2 + x\sin\theta_2)},$$

where

$$H_R^+ = -iZ^{-1}E_1^+(\cos\theta_1 e_x + \sin\theta_1 e_z + ie_y)$$
$$H_L^+ = iZ^{-1}E_2^+(\cos\theta_2 e_x + \sin\theta_2 e_z - ie_y)$$
$$H_1^- = -iZ^{-1}E_1^-(\sin\theta_1 e_z - \cos\theta_1 e_x + ie_y)$$
$$H_2^- = iZ^{-1}E_2^-(\sin\theta_2 e_z - \cos\theta_2 e_x - ie_y),$$

where

$$Z = (\mu/\varepsilon)^{1/2}[1 + (\mu/\varepsilon)\beta^2]^{-1/2}.$$

6.A.2 Appendix to Section 6.2.2 (1)

Derivation of Equation (6.43):

Paying attention to $H_i = H_0$, replace H_i of formula (6.39) with H_0,

$$j\omega m_f = \gamma(M_0 \times h_f + m_f \times H_0). \tag{6.A.1}$$

From the equation, delete the third term of Eq. (6.41):

$$m_f \times H_0 = \frac{\gamma(H_0 \cdot M_0)h - \gamma(H_0 \cdot h)M_0 - \gamma(H_0 \cdot H_0)m_f}{j\omega}. \tag{6.A.2}$$

Substitute this equation into the second term on the right side of Eq. (6.A.2),

$$j\omega m_f = \gamma(M_0 \times h) + \frac{\gamma^2(H_0 \cdot M_0) - \gamma^2(H_0 \cdot h)M_0 - \gamma^2(H_0 \cdot H_0)m_f}{j\omega}. \tag{6.A.3}$$

Here, when the final term of Eq. (6.A.3) is transformed to $(H_0 \cdot H_0)m_f = H_0^2 m_f$ and the relation of $\omega_0 = \gamma H_0$ is further applied, the expression (6.A.3) can be expressed by the following expression.

Where $\omega_0 = \gamma H_0$ is the angular velocity during free precession.

$$(\omega_0^2 - \omega^2)m_f = j\omega\gamma(M_0 \times h) + \gamma^2(H_0 \cdot M_0)h - \gamma^2(H_0 \cdot h)M_0$$

$$\therefore m_f = \frac{1}{\omega_0^2 - \omega^2}\{j\omega\gamma(M_0 \times h) + \gamma^2(H_0 \cdot M_0)h - \gamma^2(H_0 \cdot h)M\}$$

As is clear from this equation, m_f has a singular point when $\omega = \omega_0 = \gamma H_0$ is held.

6.A.3   Appendix to Section 6.2.2 (2)

Derivations of the equation of magnetic susceptibility of positive and negative circularly polarized waves:

Here, let us explain that the circularly polarized magnetic susceptibility is represented by the form of Eq. (6.64). The magnetization and the magnetic field are given by the following formulas from Eqs (6.53) and (6.54) in the text:

$$M = kM_0 + m_f e^{j\omega t} \tag{6.53}$$

$$H = kH_0 + he^{j\omega t}. \tag{6.54}$$

Here, since $m_z = 0$ and $h_z = 0$ in these equations,

$$m_f = im_x + jm_y \tag{6.A.4}$$

$$h = ih_x + jh_{yz}. \tag{6.A.5}$$

From Eq. (6.47) in the text,

$$m_x = \chi_{xx}h_x + \chi_{xy}h_y \tag{6.A.6}$$

$$m_y = \chi_{yx}h_x + \chi_{yy}h_y. \tag{6.A.7}$$

Substituting formulas (6.A.6) and (6.A.7) into formula (6.A.4),

$$m_f = (\chi_{xx}h_x + \chi_{xy}h_y)i + (\chi_{yx}h_x + \chi_{yy}h_y)j \tag{6.A.8}$$

Now, under these relational expressions, let us consider circularly polarized waves.

(a) *Case of positive circularly polarized wave (right circular polarized wave)* h^+:
In this case, since the relationship of $h_y = -jh_x$ is satisfied for the positive circularly polarized wave, the formula (6.A.8) is transformed into the

following equation:

$$m_f^+ = (\chi_{xx}h_x + \chi_{xy}h_y)\boldsymbol{i} + (\chi_{yx}h_x + \chi_{yy}h_y)\boldsymbol{j}$$
$$= (\chi_{xx}h_x - j\chi_{xy}h_x)\boldsymbol{i} + (\chi_{yx}h_x - j\chi_{yy}hx)\boldsymbol{j}$$
$$= (\chi_{xx} - j\chi_{xy})(h_x\boldsymbol{i} + h_y\boldsymbol{j}) = (\chi_{xx} - j\chi_{xy})\boldsymbol{h}^+.$$

Here, the relationship between $\chi_{xx} = \chi_{yy}$ and $\chi_{yx} = -\chi_{xy}$ is used. Hence, the positive circularly polarized magnetic susceptibility is given by the following equation:

$$\chi_+ = \chi_{xx} - j\chi_{xy}.$$

(b) *Case of negative circular polarization (left circular polarization)* $\boldsymbol{h}^-$:
Likewise, in the case of negative circularly polarized waves, Eq. (6.A.8) is modified using the relationship of $h_y = jh_x$:

$$m_f^- = (\chi_{xx}h_x + j\chi_{xy}h_x)\boldsymbol{i} + (\chi_{yx}h_x + j\chi_{yy}h_x)\boldsymbol{j}$$
$$= (\chi_{xx} + j\chi_{xy})(h_x\boldsymbol{i} + h_y\boldsymbol{j})$$
$$= (\chi_{xx} + j\chi_{xy})\boldsymbol{h}^-.$$

Therefore, the negative circularly polarized magnetic susceptibility is given by the following equation:

$$\chi_- = \chi_{xx} + j\chi_{xy}.$$

6.A.4   Appendix to Section 6.3.1 (1)

Derivation of expression (6.82):
  From Equation (6.80),

$$\boldsymbol{i}_z \times \nabla_t H_z = \boldsymbol{i}_z \times \gamma H_t + j\omega\varepsilon E_z\boldsymbol{i}_z - j\omega\varepsilon(E_t + E_z\boldsymbol{i}_z)$$
$$= \boldsymbol{i}_z \times \gamma H_t - j\omega\varepsilon E_t. \tag{6.80}$$

If the outer product of $\boldsymbol{i}_z$ is applied to both sides of the expression (6.80),

$$\boldsymbol{i}_z \times (\boldsymbol{i}_z \times \gamma H_t) = \boldsymbol{i}_z \times (\boldsymbol{i}_z \times \nabla_t H_z + j\omega\varepsilon E_t). \tag{6.A.9}$$

Using the vector triple product formula to this left side expression,

$$\boldsymbol{i}_z \times (\boldsymbol{i}_z \times \gamma H_t) = (\gamma H_t \cdot \boldsymbol{i}_z)\boldsymbol{i}_z - (\boldsymbol{i}_z \cdot \boldsymbol{i}_z)\gamma H_t$$
$$= -\gamma H_t. \tag{6.A.10}$$

The right side of Eq. (6.A.9),

$$\boldsymbol{i}_z \times (\boldsymbol{i}_z \times \nabla_t H_z + j\omega\varepsilon E_t) = \boldsymbol{i}_z \times (\boldsymbol{i}_z \times \nabla_t H_z) + \boldsymbol{i}_z \times j\omega\varepsilon E_t$$
$$= (\nabla_t H_z \cdot \boldsymbol{i}_z)\boldsymbol{i}_z - \nabla_t H_z + \boldsymbol{i}_z \times j\omega\varepsilon E_t$$
$$= -\nabla_t H_z + \boldsymbol{i}_z \times j\omega\varepsilon E_t. \tag{6.A.11}$$

Therefore, by equating both formulas (6.A.10) and (6.A.11),

$$-\gamma H_t = -\nabla_t H_z + i_z \times j\omega\varepsilon E_t$$

$$\therefore H_t = \frac{1}{\gamma}(\nabla_t H_z - i_z \times j\omega\varepsilon E_t). \tag{6.A.12}$$

Substituting this expression (6.A.12) into the left side of the expression (6.79),

$$i_z \times \gamma E_t + j\omega\mu H_t - \omega\mu' \times H_t$$

$$= i_z \times \gamma E_t + j\omega\mu\frac{1}{\gamma}(\nabla_t H_z - i_z \times j\omega\varepsilon E_t)$$

$$-\omega\mu' i_z \frac{1}{\gamma}(\nabla_t H_z - i_z \times j\omega\varepsilon E_t)$$

$$= i_z \times \gamma E_t + j\omega\mu\frac{1}{\gamma}\nabla_t H_z + i_z \times \omega^2\varepsilon\mu\frac{1}{\gamma}E_t$$

$$-\omega\mu' i_z \times \frac{1}{\gamma}\nabla_t H_z + \omega\mu' i_z \times (i_z \times j\omega\varepsilon E_t)$$

$$= i_z \times \nabla_t E_z. \tag{6.A.13}$$

By rearranging Eq. (6.A.13)

$$i_z \times K^2 - jk'^2 E_t$$

$$= \gamma i_z \times \nabla_t E_z + \omega\mu' i_z \times \nabla_t H_z - j\omega\omega\nabla_t H_z. \tag{6.A.14}$$

Furthermore, taking the outer cross product of i_z and multiplying K^2 to both sides,

$$K^4 E_t + jk'^2 K^2 i_z \times E_t$$

$$= K^2(\gamma\nabla_t E_z + \omega\mu'\nabla_t H_z + j\omega\mu i_z \times \nabla_t H_z). \tag{6.A.15}$$

By multiplying both sides of Eq. (6.A.14) by P and rearranging it,

$$k'^4 E_t + jK^2 k'^2 i_z \times E_t$$

$$= k'^2(j\gamma i_z \times \nabla_t E_z + j\omega\mu' \times \nabla_t E_z + \omega\mu\nabla_t H_z). \tag{6.A.16}$$

From the expressions (6.A.15) and (6.A.16), the following Eq. (6.82) is obtained:

$$(K^4 - k'^4)E_t = \nabla_t(K^2\gamma E_z + \omega\gamma^2\mu H_z)$$

$$+ ji_z \times \nabla_t\{\omega(K^2\mu - k'^2\mu')H_z - \gamma k'^2 E_z\}, \tag{6.82}$$

6.A.5 Appendix to Section 6.3.1 (2)

Derivation of expression (6.87):

Taking the divergence of both sides of the formulas (6.82) and (6.83),

$$
\begin{aligned}
\nabla_t \cdot (K^4 &- k'^4)\boldsymbol{E}_t \\
&= \nabla_t \cdot \nabla_t (K^2 \gamma E_z + \omega \gamma^2 \mu H_z) \\
&\quad + \nabla_t \cdot [j\boldsymbol{i}_z \times \nabla_t \{\omega (k^2 \mu - k'^2 \mu') H_z - \gamma k'^2 E_z \}] \\
&= \nabla_t^2 (K^2 \gamma E_z + \omega \gamma^2 \mu' H_z) \\
&\quad + \nabla_t \{\omega (k^2 \mu - k'^2 \mu') H_z - \gamma k'^2 E_z \} \cdot (\nabla_t \times \gamma \boldsymbol{i}_z) \\
&\quad - j\boldsymbol{i}_z \cdot [\nabla_t \times \nabla_t \{\omega (k^2 \mu - k'^2 \mu') H_z - \gamma k'^2 E_z \}] \\
&= \nabla_t (K^2 \gamma E_z + \omega \gamma^2 \mu' H_z)
\end{aligned} \tag{6.A.17}
$$

$$
\begin{aligned}
\nabla_t \cdot (K^4 &- k'^4)\boldsymbol{H}_t \\
&= \nabla_t \cdot \nabla_t (K^2 \gamma H_z + k'^2 \omega \varepsilon E_z) - \nabla_t \cdot \{j\boldsymbol{i}_z \times \nabla_t (K^2 \omega \varepsilon E_z + k'^2 \gamma H_z)^2 \} \\
&= \nabla_t^2 (K^2 \gamma H_z + k'^2 \omega \varepsilon E_z) - \nabla_t (K^2 \omega \varepsilon E_z + k'^2 \gamma H_z) \cdot (\nabla_t \times \gamma \boldsymbol{i}_z) \\
&\quad + j\boldsymbol{i}_z \cdot \{\nabla_t \times \nabla_t (K^2 \omega \varepsilon E_z + k'^2 \gamma H_z) \} \\
&= \nabla_t^2 (K^2 \gamma H_z + k'^2 \omega \varepsilon E_z).
\end{aligned} \tag{6.A.18}
$$

Here, from the divergence equations $\nabla \cdot \boldsymbol{E} e^{\gamma z} = 0$ and $\nabla \cdot \boldsymbol{B} e^{\gamma z} = 0$, there is a relationship with the following equation:

$$
\nabla_t \cdot \boldsymbol{E}_t = -\gamma E_z \tag{6.A.19}
$$

$$
\nabla_t \cdot \boldsymbol{B}_t = -\gamma B_z \tag{6.A.20}
$$

$$
\nabla_t \cdot \boldsymbol{H}_t = -\gamma \frac{\mu_z}{\mu} H_z - \omega \varepsilon \frac{\mu'}{\mu} E_z. \tag{6.A.21}
$$

The left side of the expression (6.A.17) is given by using the relationship of the expression (6.A.19).

$$
\begin{aligned}
\nabla_t \cdot (K^4 &- k'^4)\boldsymbol{E}_t \\
&= \boldsymbol{E}_t \cdot \nabla_t (K^4 - k'^4) + (K^4 - k'^4)\nabla_t \cdot \boldsymbol{E}_t \\
&= (K^4 - k'^4)\nabla_t \cdot \boldsymbol{E}_t \\
&= -(K^4 - k'^4)\gamma E_z.
\end{aligned} \tag{6.A.22}
$$

The right side of Eq. (6.A.17):

$$
\nabla_t^2 (K^2 \gamma E_z + \omega \gamma^2 \mu' H_z) = K^2 \gamma \nabla_t^2 E_z + \omega \gamma^2 \mu' \nabla_t^2 H_z. \tag{6.A.23}
$$

Therefore, Eq. (6.A.17) is expressed by the following equation:

$$
K^2 \gamma \nabla_t^2 E_z + \omega \gamma^2 \mu' \nabla_t^2 H_z + (K^4 - k'^4)\gamma E_z = 0. \tag{6.A.24}
$$

Similarly, the left side of the expression (6.A.18) can be derived using the relationship of the expression (6.A.21):

$$
\begin{aligned}
\nabla_t \cdot (K^4 - k'^4) \boldsymbol{H}_t \\
= (K^4 - k'^4) \nabla_t \cdot \boldsymbol{H}_t \\
= (K^4 - k'^4) \left(-\gamma \frac{\mu_z}{\mu} H_z - \omega\varepsilon \frac{\mu'}{\mu} E_z \right).
\end{aligned}
\tag{6.A.25}
$$

Accordingly, Eq. (6.A.18),

$$
K^2 \gamma H_z + k'^2 \omega\varepsilon \nabla_t^2 E_z + \left(\gamma \frac{\mu_z}{\mu} H_z + \omega\varepsilon \frac{\mu'}{\mu} E_z \right) = 0.
\tag{6.A.26}
$$

From Eq. (6.A.25),

$$
\omega\gamma^2 \mu' \nabla_t^2 H_z = -K^2 \gamma \nabla_t^2 E_z - (K^4 - k'^4) \gamma E_z
\tag{6.A.27}
$$

$$
\begin{aligned}
\therefore \nabla_t^2 H_z &= -K^2 \frac{\gamma}{\omega\gamma^2 \mu'} \nabla_t^2 E_z - \frac{\gamma}{\omega\gamma^2 \mu'} (K^4 - k'^4) \gamma E_z \\
&= -\frac{K^2}{\omega\gamma \mu'} \nabla_t^2 E_z - \frac{1}{\omega\gamma \mu'} (K^4 - k'^4) E_z.
\end{aligned}
\tag{6.A.28}
$$

Substituting Eq. (6.A.28) into Eq. (6.A.26),

$$
\begin{aligned}
(k'^2 \omega\varepsilon) \nabla_t^2 E_z + K^2 \gamma \left\{ -\frac{K^2}{\omega\gamma \mu'} \nabla_t^2 E_z - \frac{1}{\omega\gamma \mu'} (K^4 - k'^4) E_z \right\} \\
+ (K^4 - k'^4) \left(\gamma \frac{\mu_z}{\mu} H_z + \omega\varepsilon \frac{\mu'}{\mu} E_z \right) \\
= \left(k' \omega\varepsilon - \frac{K^4 \gamma}{\omega\gamma \mu'} \right) \nabla_t^2 E_z + (K^4 - k'^4) \left(\frac{\omega\gamma \mu'}{\mu} - \frac{K^2 \gamma}{\omega\gamma \mu'} \right) E_z \\
+ (K^4 - k'^4) \gamma \frac{\mu_z}{\mu} H_z = 0.
\end{aligned}
\tag{6.A.29}
$$

If both sides of the formula (6.A.29) are divided by $k'^2 \omega\varepsilon - K^4 \gamma / (\omega\gamma \mu')$,

$$
\nabla_t^2 E_z + a E_z + b H_z = 0,
\tag{6.87}
$$

where $a = K^2 - \frac{k'^2 \mu'}{\mu}$, $b = -\frac{\omega\gamma \mu' \mu_z}{\mu}$.

6.A.6 Appendix to Section 6.3.1 (3)

Derivation of expression (6.107):
   The roots of the quadratic equation in the expression (6.106) are S_1, S_2, respectively.

Using the relationship between roots and coefficients in this quadratic equation, Eq. (6.107) can be presented as follows:

$$S_1 S_2 = ac - bd$$

$$= \left(K^2 - \frac{k'^2 \mu'}{\mu} \right) \left(\frac{K^2 \mu_z}{\mu} \right) + \left(\frac{\omega \gamma \mu' \mu_z}{\mu} \right) \left(\frac{\omega \gamma \varepsilon \mu'}{\mu} \right)$$

$$= \frac{\mu_z}{\mu} (K'^4 - k'^4)$$

$$\therefore K'^4 - k'^4 = \frac{\mu}{\mu_z} S_1 S_2,$$

where a, b, c, and d are constants given in Eqs. (6.87) and (6.88).

References

1 Engheta, N. and Jaggard, D.L. (1988). Electromagnetic chirality and its applications. *IEEE AP-S News let.* 30 (5): 6–12.

2 Bassiri, S., Papas, C.H., and Engheta, N. (1988). Electromagnetic wave propagation through a dielectric-chiral interface and through a chiral slab. *J. Opt., Soc. Am. A.* 5 (9): 1650–1659.

3 Jaggard, D.L., Engheta, N., Kowarz, M.W. et al. (1989). Periodic chiral structure. *IEEE Trans. Antennas Prpopag.* 37 (11): 1667–1652.

4 Cuire, T. and Varadan, V.V. (1990). Influence of Chirality on the Reflection of EM Waves by planar dielectric slabs. *IEEE Trans. Electromagn. Compat.* 32 (4).

5 Tanaka, M. and Kusunoki, A. (1993). Scattering characteristics of stratified chiral slab. *IEICE Trans. Electron.* 76-C, 10: 1443–1448.

6 Kotsuka, Y. and Wakita, H. (1995). Analysis on matching characteristics of a multi-layered chiral medium mixed with a magnetic material. *IEICE Tech Rep. EMCJ* 95-59: 17–22.

7 Collin, R.E. (1966). *Foundations for microwave engineering.* McGraw-Hill Book Company.

8 Kotsuka, Y. (1995). *Electromagnetic Wave Analysis,* 365–396. Tokyo, Japan: Corona Publishing Company CO., LTD.

9 Kales, M.L. (1953). Mode in wave guides containing ferrite. *J. Appl. Phys* 26 (5).

10 Kotsuak, Y. (1983). Rigorous analysis of microwave attenuation characteristics in a coaxial waveguide with anisotropic ferrite medium. In: *1983-URSI Symposium on Electromagnetic Theory, Santiago DE Compostwela (Spain),* 679–682.

11 Y. Kotsuak. "Propagation characteristics in a coaxial waveguide filled with a weekly magnetized ferrite, in URSI Radio Sci. Meet. Virginia Tech, vol. UB07-6, Blacksburg, VA. P.123, June, 1987.

12 Suhl, H. and Walker, L.R. (1956). Topics in guided wave propagation through gyromagnetic medium. *Part III, BSTJ* 3: 1168–1196.

13 Kotsuka, Y. and Yamazaki, H. (2000). Fundamental investigation on a weekly magnetized ferrite absorber. *IEEE Trans. EMC* 62 (2): 116–124.

14 Collin, R.E. (1960). *Field Theory of Guided Waves*, 509–551. McGraw-Hill Book Company.

15 Kotsuka, Y. (2010). Metamaterial overview: Investigation of the application of metamaterial technology to next generation communications. *J. IEICE* 93 (6): 636–639.

16 Veselago, V.G. (1968). The electrodynamics of substances with simultaneously negative values of ε and μ. *Sov. Phys. Usp.* 10 (6): 509–516.

17 Pendry, J.B., Holden, A.J., Stewart, W.J., and Voungs, I. (1996). Extremely Low Frequency Plasmons in Metallic Mesostructures. *Phys. Rev. Lett.* 25: 76.

18 Pendry, J.B., Holden, A.J., Robbins, D.J., and Stewart, W.J. (1999). Magnetism form conductors and enhanced nonlinear phenomena. *IEEE Trans. Microwave Theory Tech.* 67 (11): 2075–1086.

19 Smith, D.R., Padilla, W.J., Vier, D.C. et al. (2000). Composite medium with simultaneously negative permeability and permittivity. *Phys. Rev. Lett.* 86 (18): 6186–6187.

20 Shelby, R.A., Smith, D.R., and Shultz, S. (2001). Experimental verification of a negative index of refraction. *Science* 292: 77–79.

21 Caloz, C. and Itoh, T. (2002). Application of the transmission line theory of left-handed materials to the realization of a microstrip LH transmission line. In: *Proceedings of the IEEE-AP-SUNC/URSI National Radio Science Meeting*, vol. 2, 612–615. San Antonio, TX.

22 Iyer, A.K. and Eleftheriades, G.V. (2002). Negative refractive index metamaterials support-supporting 2-D waves. In: *IEEE MTT International, Symposium*, vol. 2, 612–615. Seattle, WA.

23 Oliner, A.A. (2002). A periodic-structure negative-refractive–index medium without resonant element. In: *URSI Digest, IEEE-APS USNC?URSI National Radio Science Meeting, San Antonio, TX*, 61.

24 Caloz, C. and Itoh, T. (2006). *Electromagnetic Metamaterials*, 8. Wiley-Interscience.

25 Sanada, A., Caloz, C., and Itoh, T. (2004). Characteristics of the composite right/left-handed transmission lines. *IEEE Microwave and Wireless Component Letters* 14 (2): 68–70.

26 Sanada, A., Caloz, C., and Itoh, T. (2004). Planar distributed structures with negative refractive index. *IEEE Trans. on Microwave Theory and Techniques* 52 (4): 1252–1263.

27 Eleftheriades, G.V. and Balmain, K.G. (2005). *Negative–Refraction Metamaterials*. IEEE Press/Wiley.

28 Ishimaru, A., Lee, S.W., Kuga, Y., and Jandhyala, V. (2003). General constitutive relations for metamaterials based on the quasi-static Lorentz theory. *IEEE Trans. Antenas Propagat.* 51 (10): 2550–2557.

29 Lewin, L. (1967). The electrical constants of a material loaded with spherical particles. *Proc. Inst, Elec. Eng.* 96: 65–68.

30 Vendick, O.G. and Gashinova, M.S. (1998). Artificial double negative media composed by two different dielectric sphere lattice embedded in a dielectric matrix. In: *Proceedings of European Microwave Conference, Amsterdam, The Netherland*, 6785–6809.

31 Yen, T.J., Padilla, W.J. et al. (2006). Terahertz magnetic response form artificial material. *Science* 303: 1696–1696.

32 Moser, H.O., Casse, B.D.F. et al. (2005). Terahertz response of a microfabricated rod-split-ring-resonator electromagnetic material. *Phys. Rev. Tett.* 96: 1–6.

33 Bilotti, F., Toscano, A., Alici, K.B. et al. (2011). Design of miniaturized narrow band absorbers based on resonant-magnetic, inclusions. *IEEE Trans. EMC* 53 (1): 63–71.

34 Landy, N.I., Sajuyigbe, S., Mock, J.J. et al. (2008). Perfect metamaterial absorber. *Phys. Rev. Lett.* 100: 207602.

7

Measurement Methods on EM-Wave Absorbers

In order to deepen our knowledge of the electromagnetic (EM)-wave absorber design, it becomes important to accurately know the material characteristics and the characteristics of the absorber itself.

In this chapter, after describing the measurement methods of material constants, the measurement methods of EM-wave absorber characteristics are described.

In particular, we intend here to explain in detail the measurement methods of material constants, including conventional methods together with analytical methods. In Section 7.1, as the methods of measuring the material constants for wave-absorbing materials, the methods of using a rectangle waveguide, coaxial waveguide, and microwave resonators are described. First, measurement methods based on the standing-wave method using the rectangular waveguide and coaxial waveguide are studied. Second, in the microwave resonator methods for cylindrical cavities, the measurement methods of dielectric constant and conductivity in the lossless case, complex permittivity, and complex permeability are investigated theoretically. In Section 7.2, as a method for measuring the EM-wave absorber characteristics, the method of TEM mode transmission lines such as coaxial waveguides, strip lines, and TEM cells are examined first. Then, the waveguide method, as a special case, is described, along with space standing-wave methods.

7.1   Material Constant Measurement Methods

To evaluate (design) the EM-wave absorber characteristics, it is necessary to accurately know the material constants. Needless to say, once the values of the permittivity and permeability are determined accurately, the rigorous evaluations for the EM-wave absorption characteristics become possible. Recently, for these material constant measurements, the method of determining both the permittivity and permeability values was introduced by accurately measuring the S-parameters using a network analyzer. However, usually, because the

Electromagnetic Wave Absorbers: Detailed Theories and Applications, First Edition. Youji Kotsuka.
© 2019 John Wiley & Sons, Inc. Published 2019 by John Wiley & Sons, Inc.

measurement methods of the material depend on the material of the EM-wave absorber and the application frequencies, they cannot be determined simply. Thus, various measurement methods should be conducted in each particular case. Currently, the required measurement frequency of the EM-wave absorber is also extended to terahertz regions. In general, measurements of the dielectric constant and magnetic permeability in the high-frequency band were conducted by the standing-wave method, the transmission line method, the resonance method (perturbation method), the free space method, and the like [1]. In the following, let us first describe the material constant measurement method, particularly from the viewpoint of the basic theory. Then, methods for measuring the EM-wave absorber characteristics are introduced.

7.1.1 Standing-Wave Method

7.1.1.1 Case of Using Waveguide

By inserting the measurement sample at the front surface of the waveguide with a short-circuited termination and performing the standing-wave measurement, the input impedance value looking at the measurement sample can be measured, and the material constants of the sample can be calculated.

Let us first consider the case of inserting the sample of thickness d at the end of the waveguide composed of a conductor plate and measuring standing waves based on a TE_{10} wave (fundamental wave), as shown in Figure 7.1a. In the case of the TE_{10} mode propagation in the waveguide, we now express the

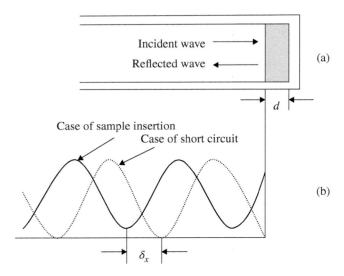

Figure 7.1 Standing-wave measurement using a waveguide. (a) Waveguide loaded with sample. (b) Standing wave distribution.

propagation constants of the air medium and lossy measurement sample using γ_0 and γ_1, respectively. These propagation constants are given by the following equations:

$$\gamma_0 = \sqrt{k_c^2 - k^2} = \sqrt{\left(\frac{2\pi}{\lambda_c}\right)^2 - \left(\frac{2\pi}{\lambda_0}\right)^2} = \frac{2\pi}{\lambda_0}\sqrt{\left(\frac{\lambda_0}{\lambda_c}\right)^2 - 1} = j\beta_0 = j\frac{2\pi}{\lambda_g},$$
$$(7.1)$$

$$\gamma_1 = \sqrt{k_c^2 - \omega^2\varepsilon\mu} = \sqrt{\left(\frac{2\pi}{\lambda_c}\right)^2 - \left(\frac{2\pi}{\lambda_0}\right)^2\varepsilon_s\mu_s} = \frac{2\pi}{\lambda_0}\sqrt{\left(\frac{\lambda_0}{\lambda_c}\right)^2 - \varepsilon_s\mu_s}.$$
$$(7.2)$$

Here, k_c is represented by $2\pi/\lambda_c$ as the eigenvalues associated with the cutoff wavelength λ_c, λ_0 is the wavelength in free space, k is the phase constant in free space, λ_g is the guide wavelength, and ε_s and μ_s represent the relative permittivity and the relative permeability, respectively. Since the measurement of a dielectric material sample is considered here, the relative permeability value is defined as $\mu_s = 1$. When the air medium is expressed using the dielectric constant ε_0 and the magnetic permeability μ_0 of the vacuum medium, the wave impedance Z_{h0} in the air is represented by the following equation:

$$Z_{h0} = \frac{j\omega\mu_0}{\gamma_0}.$$
$$(7.3)$$

Also, the wave impedance Z_{h1} in the sample with losses of a dielectric material is

$$Z_{h1} = \frac{j\omega\mu_0}{\gamma_1}.$$
$$(7.4)$$

Here, the wave impedance Z_{h1} of the sample medium normalized by the Z_{h0} in the air medium can be represented by z_s, as follows:

$$z_s = \frac{Z_{h1}}{Z_{h0}} = \frac{\gamma_0}{\gamma_1} = \frac{\sqrt{\left(\frac{\lambda_0}{\lambda_c}\right)^2 - 1}}{\sqrt{\left(\frac{\lambda_0}{\lambda_c}\right)^2 - \varepsilon_s}}.$$
$$(7.5)$$

Next, let us denote the impedance Z_{in} that looks at the sample front surface by normalized impedance z_{in}.

$$z_{in} = z_s \tanh \gamma_1 d.$$
$$(7.6)$$

Substituting the expression $z_s = \gamma_0/\gamma_1$ in Eq. (7.5) and the $\gamma_0 = j\beta_0$ in Eq. (7.1) into (7.6) and dividing by d, the following expression can be derived:

$$\frac{z_{in}}{j\beta_0 d} = \frac{\tanh \gamma_1 d}{\gamma_1 d}.$$
$$(7.7)$$

Now, after measuring a standing-wave distribution in the case of a short-circuited terminal and in the case of inserting the sample, the distance between the minimum electric field points is denoted by δ_x, as shown in Figure 7.1b,

$$\delta_x = |x_1 - x_0|. \tag{7.8}$$

Next, letting ρ be the standing-wave ratio at the time of inserting the sample, z_{in} is expressed via the standing-wave ratio as follows:

$$z_{in} = \frac{1 - j\rho \tanh \frac{2\pi\delta_x}{\lambda_g}}{\rho - j \tanh \frac{2\pi\delta_x}{\lambda_g}}. \tag{7.9}$$

Therefore, z_{in} is determined from Eq. (7.9). Substituting this expression into Eq. (7.7), and solving Eq. (7.7) for $\gamma_1 d$, since $\beta_0 (= 2\pi/\lambda_g)$ and d are known, it becomes possible to determine the propagation constant $\gamma_1 d$ in the sample. Notice here that β_0 and d are the known values. However, since Eq. (7.7), generally, is a so-called transcendental equation, it cannot be solved easily for $\gamma_1 d$. Conventionally, for this type of equation, a method of obtaining the solution Z by referring to a previously prepared analysis chart by expressing Eq. (7.7) in the form of the following has been taken.

$$W = \frac{\tanh Z}{Z}. \tag{7.10}$$

here, $W = z_{in}/j\beta d$, $Z = \gamma_1 d$.

However, recently, it became possible to find an accurate solution for this kind of transcendental equation in numerical analysis by a computer, for example, using Newton's method. If the propagation constant γ_1 in the sample medium can be obtained as a solution by computer analysis, the real and imaginary parts of the dielectric constant of the sample medium can be expressed as follows.

Now, if denoting the final solution, namely, the propagation constants in the sample medium, in the following form:

$$\gamma_1 = \alpha + j\beta. \tag{7.11}$$

Then, from Eq. (7.2), the following equation is given:

$$\alpha + j\beta = \left(\frac{2\pi}{\lambda_0}\right) \sqrt{\left(\frac{\lambda}{\lambda_c}\right)^2 - (\varepsilon_s' - j\varepsilon_s'')}. \tag{7.12}$$

Here, squaring both sides of Eq. (7.12) and equalizing the real and imaginary parts of both sides, the permittivity of the test sample can be derived:

$$\varepsilon_s' = \left(\frac{\lambda}{\lambda_c}\right)^2 + \frac{\beta^2 - \alpha^2}{(2\pi/\lambda)^2}, \tag{7.13}$$

$$\varepsilon_s'' = \frac{2\alpha\beta}{(2\pi/\lambda)^2}. \tag{7.14}$$

Also, $\tan\delta$ can be expressed in the following expression:

$$\tan\delta = \frac{2\alpha\beta}{(2\pi/\lambda_c)^2 + \beta^2 - \alpha^2}.$$

7.1.1.2 Method of Using Coaxial Waveguides

If the measurement sample can be made in the shape of a circular donut, the permittivity and permeability values can be obtained by the standing-wave method using a coaxial waveguide.

Methods using a coaxial waveguide are classified as follows:

(a) The method of measuring the standing wave when one measuring sample is placed in front of the shorting terminal end of the coaxial waveguide, as shown in Figure 7.2a;

(b) The method of measuring the standing wave when two measuring samples are placed in front of the shorting terminal end of the coaxial waveguide by piling up on top of each other, as shown in Figure 7.2a,b;

(c) A method for obtaining the permittivity and permeability at the same time by placing the measurement specimen in a position separated by $1/4\lambda$ from the front end of the coaxial waveguide terminal end and the coaxial waveguide end, as shown in Figure 7.3a,b.

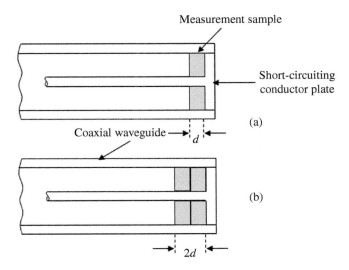

Figure 7.2 Material constant measurement with a coaxial tube. (a) When one measurement specimen is loaded at the end of the coaxial waveguide. (b) In the case of two specimens.

First, let us describe the method of (a) measuring the standing wave when a measuring single sample is placed in front of the shorting terminal end of the coaxial waveguide.

This method is effective for measuring the material constant of a magnetic material typified by ferrite.

The input impedance looks into the sample from its surface, when the samples are mounted on the front of the coaxial waveguide termination,

$$z_{in} = \sqrt{\frac{\dot{\mu}_r}{\dot{\varepsilon}_r}} \tanh\left(j\frac{2\pi}{\lambda}\sqrt{\dot{\varepsilon}_r\dot{\mu}_r}d\right) = \sqrt{\frac{\dot{\mu}_r}{\dot{\varepsilon}_r}}\tanh W. \tag{7.15}$$

This formula represents the input impedance normalized by the free space impedance $z_0 = \sqrt{\mu_0/\varepsilon_0}$. Here, $W = j2\pi\sqrt{\dot{\varepsilon}_r\dot{\mu}_r}d/\lambda$, λ is the wavelength, d is the sample thickness, and $\dot{\varepsilon}_r$ and $\dot{\mu}_r$ are the complex relative permittivity and permeability, respectively.

The left-hand side z_{in} of Eq. (7.15) is determined by measuring the standing wave. If $\dot{\varepsilon}_r$ of the right side is a known value, Eq. (7.15), which is called the transcendental equation in mathematics, can be numerically analyzed on a computer.

In particular, in the case of a low frequency, if the sample thickness is thin and the absolute values of $\dot{\varepsilon}_r$ and $\dot{\mu}_r$ do not take large values, the relation $|W| \ll 1$ can be held. Since in this case the $\tanh W \cong W$ is satisfied, Eq. (7.15) can be easily solved, and $\dot{\mu}_r$ is approximated by the following equation:

$$\dot{\mu}_r \cong -j\frac{\lambda}{2\pi d}z_{in}. \tag{7.16}$$

However, in many cases, it is necessary to know both the $\dot{\varepsilon}_r$ and $\dot{\mu}_r$ values.

In this case, we have to take the procedure of case (b), as pointed out earlier. This case can be solved by the method shown in Figure 7.2a,b.

That is, to do this, one first has to determine $z_{in}(= z_{in1})$ with a sample thickness d, utilizing the standing-wave measurement, and then z_{in2} must be determined with the sample thickness $2d$. After this, it is necessary to solve these equations for z_{in1} and z_{in2} to determine the values of $\dot{\varepsilon}_r$ and $\dot{\mu}_r$. Let us denote here these normalized impedance expressions by omitting the subscript "in" for simplicity:

$$z_1 = \sqrt{\frac{\dot{\mu}_r}{\dot{\varepsilon}_r}}\tanh\left(j\frac{2\pi}{\lambda}\sqrt{\dot{\varepsilon}_r\dot{\mu}_r}d\right),$$

$$z_2 = \sqrt{\frac{\dot{\mu}_r}{\dot{\varepsilon}_r}}\tanh\left(j\frac{2\pi}{\lambda}\sqrt{\dot{\varepsilon}_r\dot{\mu}_r}2d\right).$$

Solving these two equations for $\dot{\mu}_r$ and $\dot{\varepsilon}_r$, the following expressions yield after some calculations are performed:

$$\dot{\mu}_r = -j \frac{\lambda z_1}{4\pi d} \frac{1}{\sqrt{2z_1/z_2 - 1}} \ln\left(\frac{1 + \sqrt{2z_1/z_2 - 1}}{1 - \sqrt{2z_1/z_2 - 1}}\right), \tag{7.17}$$

$$\dot{\varepsilon}_r = \frac{\dot{\mu}_r}{z_1^2}\left(\frac{2z_1}{z_2}\right). \tag{7.18}$$

Consequently, the complex-valued magnetic permeability and complex-valued permittivity can be calculated for known z_1 and z_2.

Next, let us introduce the case of (c). This is a method of obtaining the permittivity and permeability simultaneously by placing a measurement specimen in a position separated by $1/4\lambda$ from the front end of the coaxial waveguide terminal end and the coaxial waveguide end. This case is shown in Figure 7.3a,b.

With this method, both the complex relative permittivity $\dot{\varepsilon}_r = \varepsilon'_r - j\varepsilon''_r$ and the complex relative permeability $\dot{\mu}_r = \mu'_r - j\mu''_r$ can be obtained analytically in the case of ferrite. In the case of Figure 7.3a, the normalized input impedance can be represented by z_1. Also, the case where the measurement sample is positioned at a point separated by $1/4\lambda$ from the coaxial waveguide terminal end can be expressed by z_2:

$$z_1 = \frac{Z_1}{Z_0} = \sqrt{\frac{\dot{\mu}_r}{\dot{\varepsilon}_r}} \tanh\left(j\frac{2\pi}{\lambda}\sqrt{\dot{\varepsilon}_r \dot{\mu}_r}d\right), \tag{7.19}$$

$$z_2 = \frac{Z_2}{Z_0} = \sqrt{\frac{\dot{\mu}_r}{\dot{\varepsilon}_r}} \coth\left(j\frac{2\pi}{\lambda}\sqrt{\dot{\varepsilon}_r \dot{\mu}_r}d\right). \tag{7.20}$$

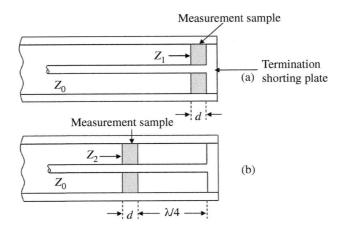

Figure 7.3 Measurements of dielectric constant and magnetic permeability with a coaxial waveguide. (a) A measurement sample is loaded at the end of the coaxial waveguide. (b) Measurement sample was loaded at a point away from the coaxial waveguide end by $1/4\lambda$.

Here, λ is the wavelength and $\dot{\varepsilon}_r = \varepsilon'_r - j\varepsilon''_r$ and $\dot{\mu}_r = \mu'_r - j\mu''_r$ represent the complex relative permittivity and relative permeability, respectively.

Taking the product of both sides of Eqs. (7.19) and (7.20), and then dividing both sides of these equations, the following relations can be derived:

$$\sqrt{\frac{\dot{\mu}_r}{\dot{\varepsilon}_r}} = \sqrt{z_1 z_2}, \tag{7.21}$$

$$\sqrt{\dot{\varepsilon}_r \dot{\mu}_r} = -j\frac{\lambda}{2\pi d}\tanh^{-1}\sqrt{\frac{z_1}{z_2}}. \tag{7.22}$$

By multiplying the left-hand sides in Eqs. (7.21) and (7.22), and the same for the right-hand sides, one obtains

$$\dot{\mu}_r = -j\frac{\lambda}{2\pi d}\sqrt{z_1 z_2}\tanh^{-1}\sqrt{\frac{z_1}{z_2}}. \tag{7.23}$$

In addition, dividing Eq. (7.22) by Eq. (7.21),

$$\dot{\varepsilon}_r = -j\frac{\lambda}{2\pi d}\frac{1}{\sqrt{z_1 z_2}}\tanh^{-1}\sqrt{\frac{z_1}{z_2}}. \tag{7.24}$$

Using z_1 and z_2 which are expressed by the following expressions:

$$z_1 = r_1 + jx_1,$$
$$z_2 = r_2 + jx_2.$$

The following relations are provided:

$$\sqrt{\frac{z_1}{z_2}} = \sqrt{x + jy}, \tag{7.25}$$

$$\sqrt{z_1 z_2} = \sqrt{x' + jy'}, \tag{7.26}$$

where

$$x = \frac{r_1 r_2 + x_1 x_2}{r_1^2 + x^2}, y = \frac{r_1 x_2 - r_2 x_1}{r_1^2 + x_1^2}, \tag{7.27}$$

$$x' = r_1 r_2 - x_1 x_2, y' = x_1 r_2 + x_2 r_1. \tag{7.28}$$

Furthermore, let us represent Eqs. (7.25) and (7.26) by the following expressions:

$$\sqrt{x + jy} = \alpha + j\beta,$$
$$\sqrt{x' + jy'} = C + jD.$$

From these equations, the following relations can be derived:

$$\alpha = \sqrt{\frac{\sqrt{x^2 + y^2} + x}{2}}, \quad \beta = \sqrt{\frac{\sqrt{x^2 + y^2} - x}{2}} \tag{7.29}$$

$$C = \sqrt{\frac{\sqrt{x'^2 + y'^2} + x'}{2}}, \quad D = \pm\sqrt{\frac{\sqrt{x'^2 + y'^2} - x'}{2}} \quad (y' \geq 0). \tag{7.30}$$

After representing again the expressions (7.25) and (7.26) using these relational expressions (7.27)–(7.30), permeability and permittivity can be calculated from Eqs. (7.23) and (7.24), respectively. That is, we finally obtain the following expressions:

$$\mu'_r = AC - BD, \quad \mu''_r = -(BC + AD) \tag{7.31}$$

$$\varepsilon'_r = \frac{AC + BD}{C^2 + D^2}, \quad \varepsilon''_r = \frac{AD - BC}{C^2 + D^2}. \tag{7.32}$$

Here,

$$A = \frac{\lambda}{2\pi d} \frac{1}{2} \left[\tan^{-1}\left(\frac{\beta}{1+\alpha}\right) + \tan^{-1}\left(\frac{\beta}{1-\alpha}\right) \right],$$

$$B = \frac{\lambda}{2\pi d} \frac{1}{2} \log \sqrt{\frac{(1-\alpha)^2 + \beta^2}{(1+\alpha)^2 + \beta^2}}.$$

This method is effective, especially when the sample thickness is small compared to the wavelength in the sample medium, although the analysis procedures are complex.

7.1.2 Cavity Resonator Method

7.1.2.1 Method of Micro-sample Insertion

As a method of determining material constant, the methods of putting a minute solid sample in a cavity resonator were devised from long ago.

In this section, the methods for obtaining material constants are introduced by measuring a small change in the "resonant frequency f_0" of the cavity and the "Q value," indicating the sharpness of the resonance. Given is a list of methods for measuring material constants below, which are considered here:

(a) The method of measuring real permittivity and conductivity,
(b) The method of complex permittivity,
(c) The method of complex permeability.

In these measurement methods, the TM_{011}, TM_{010}, and TE_{011} modes in cylindrical resonant cavities are applied to determine each material constant.

As shown in Figure 7.4, let us first consider a cavity with a volume V surrounded by a surface S made of a perfect conductor which is externally shielded. When inserting a small solid sample of volume ΔV into this cavity, the resonance frequency f_0 and the value of Q, which indicates the sharpness of the resonance in the cavity, change simultaneously. To formulate this phenomenon, the Boltzmann–Ehrenfest theorem has to be introduced. This theorem is defined as follows. In a system which periodically vibrates in a closed space, in the case where the adiabatic variation is slowly changed, the

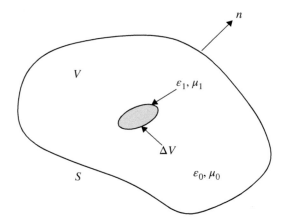

Figure 7.4 General figure of a micro-sample being put into a closed space.

ratio of the angular frequency ω_0 of the vibration to the total time-average value $\overline{W}$ of the energy contained in the system is invariant. That is

$$\overline{W}/\omega_0 = \text{invariant.} \tag{7.33}$$

This expression can be written as follows:

$$\frac{\delta\omega_0}{\omega_0} = \frac{\delta\overline{W}}{\overline{W}}. \tag{7.34}$$

Here, $\partial\omega_0$ and $\partial\overline{W}$ mean the minute change in the amounts of angular frequency and the minute change of total average energy, respectively. The applicable conditions of this theorem are limited under the condition that the adiabatic change is very slow compared with the frequency oscillation period. Also, this theorem can be applied when a small deformation of the closed surface or the appearance change of a small object comparing to the closed space has occurred.

Now, the angular frequency ω, the electromagnetic fields (E, H), and the material constants (ε, μ) are represented using a suffix 0 in the case of the initial state without the sample in the cavity, and also using a suffix 1 for the state with the sample. Therefore, the permittivity and the permeability when a measurement sample consists of the micro-volume ΔV placed in the cavity can be represented by ε_1 and μ_1, respectively, as shown in Figure 7.4. When treating the problem of electromagnetic wave, the relationship in Eq. (7.34) can be represented by the following equation:

$$\frac{\omega_1 - \omega_0}{\omega_0} = -\frac{\int_{\Delta V}(\varepsilon_1 - \varepsilon_0)E_1 \cdot E_0^* dV + \int_{\Delta V}(\mu_1 - \mu_0)H_1 \cdot H_0^* dV}{\varepsilon_0 \int_V E_0 \cdot E_0^* dV + \mu_0 \int_V H_0 \cdot H_0^* dV}. \tag{7.35}$$

(See Appendix 7.A.1.)

Furthermore, rewriting this expression as

$$\frac{\omega_1 - \omega_0}{\omega_1} \approx \frac{\delta\omega_0}{\omega_0} = -\frac{1}{4\overline{W}} \int_{\Delta V} (\Delta\varepsilon E_1 \cdot E_0^* + \Delta\mu H_0^* H_1) dV, \tag{7.36}$$

where $\Delta\varepsilon$ and $\Delta\mu$ denote $\Delta\varepsilon = \varepsilon_1 - \varepsilon_0$ and $\Delta\mu = \mu_1 - \mu_0$, respectively. E_0^* and H_0^* mean the electromagnetic fields of the complex conjugate with respect to E_0 and H_0, respectively. In the following, notice that the approximations $E_0 \cong E_1, H_0 \cong H_1$ can be applied when the measurement sample is extremely small and if the electromagnetic fields in the radial direction in sample are considered to be almost uniform before and after the sample insertion. $4\overline{W}$ in the denominator of Eq. (7.36) is the total stored energy in the cavity prior to inserting the sample. $\overline{W}$ is generally given by the following equation:

$$\overline{W} = \overline{W}_e + \overline{W}_m = 2\overline{W}_e = \frac{1}{2} \int_V \varepsilon_0 |E_0|^2 dV$$

$$= 2\overline{W}_m = \frac{1}{2} \int_V \mu_0 |H_0|^2 dV$$

$$\text{Where, } \overline{W}_e = \frac{1}{4} \int_V \varepsilon_0 |E_0|^2 dV$$

$$\overline{W}_m = \frac{1}{4} \int_V \mu_0 |H_0|^2 dV. \tag{7.37}$$

Accordingly, the denominator $4\overline{W}$ in Eq. (7.36) can be derived using the approximation $E_0 \cong E_1, H_0 \cong H_1$ and the relation of the total stored energy in the cavity in Eq. (7.37).

Now, if the cavity medium before the sample insertion is regarded as a vacuum medium, when representing its medium constants by μ_0 and ε_0, $\Delta\mu$ and $\Delta\varepsilon$ can be given by the following expressions:

$$\Delta\mu = \mu_1 - \mu_0 = \mu_0(\mu_{r1} - 1) = \mu_0 \chi_m \tag{7.38}$$

$$\Delta\varepsilon = \varepsilon_1 - \varepsilon_0 = \varepsilon_0(\varepsilon_{r1} - 1) = \varepsilon_0 \chi_e. \tag{7.39}$$

Here, $\mu_{r1}, \varepsilon_{r1}$ denote the relative permeability and relative permittivity in the measurement sample, respectively. In the case where these values denote complex relative permeability and permittivity, these material constants are represented as $\mu_{r1} = \mu'_{r1} - j\mu''_{r1}$, $\varepsilon_{r1} = \varepsilon'_{r1} - j\varepsilon''_{r1}$, respectively. χ_m and χ_e represent relative susceptibility and relative polarizability, respectively.

Representing these values with complex numbers

$$\chi_m = \chi_{mr} - j\chi_{mi}, \tag{7.40}$$

$$\chi_e = \chi_{er} - j\chi_{ei}. \tag{7.41}$$

Furthermore, when the micro-sample with the volume ΔV is placed in the resonator cavity, the complex resonant angular frequency ω_1 can be expressed in the following equation:

$$\omega_1 = \omega_r + j\omega_i = \omega_r + j(\omega_r/2Q). \tag{7.42}$$

Here, ω_1 represents the resonant complex angular frequency. ω_i satisfies a relationship $\omega_i = \omega_r/2Q$. Q is the quality factor of the resonant circuit.

Now, substituting the relationships (7.38)–(7.42) into Eq. (7.36) and separating the real part and the imaginary part, the following equations are obtained. But, notice here that the rule of approximations $E_1 \cong E_0$ and $H_1 \cong H_0$, which were pointed out in the earlier description, are now applied in the present expressions:

$$\frac{\omega_1 - \omega_0}{\omega_0} + j\frac{(\omega_r/2Q)}{\omega_0} =$$

$$-\frac{\varepsilon_0\chi_{er}|E_0|^2_{\Delta V} + \mu_0\chi_{mr}|H_0|^2_{\Delta V} - j(\varepsilon_0\chi_{ei}|E_0|^2_{\Delta V} + \mu_0\chi_{mi}|H_0|^2_{\Delta V})}{4\overline{W}}\Delta V,$$

$$\therefore \frac{\omega_r - \omega_0}{\omega_0} = \frac{\delta\omega_r}{\omega_0} = -\frac{\varepsilon_0\chi_{er}|E_0|^2_{\Delta V} + \mu_0\chi_{mr}|H_0|^2_{\Delta V}}{4\overline{W}}\Delta V \text{ (Real part)}, \tag{7.43}$$

$$\frac{\omega_r}{\omega_0}\left(\frac{1}{2Q}\right) = \delta\left(\frac{1}{2Q}\right) = \frac{\varepsilon_0\chi_{ei}|E_0|^2_{\Delta V} + \mu_0\chi_{mi}|H_0|^2_{\Delta V}}{4\overline{W}}\Delta V \text{ (Imaginary part)}. \tag{7.44}$$

Here, the symbol ΔV in the subscripts means a value related to a small solid sample.

With these relationships, it becomes possible to determine the medium constants of the sample from the measured values of $\delta\omega_r$ and $\delta(1/2Q)$ by putting the small solid sample in the cavity.

First, by loading a small solid sample into a cylindrical cavity resonator, let us show the method of determining the dielectric constant by measuring the resonant frequency in TM_{011} mode and by measuring the Q value as an example.

Generally, in the material constant measurement using this kind of cylindrical cavity:

(a) When measuring the value of the permittivity, it is necessary to select a resonance mode in which the distribution density of the electric lines of force becomes maximum at the position where the measurement sample is placed.

(b) Also, when measuring the value of the permeability, it is necessary to select a resonance mode in which the distribution density of the lines of magnetic force becomes maximum at the position where the measurement sample is placed.

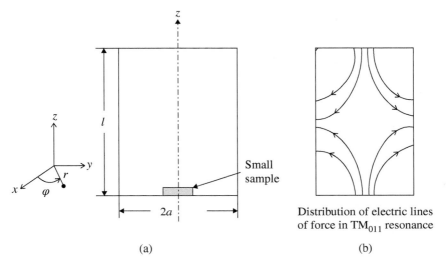

(a)   (b)

Figure 7.5 Cylindrical cavity resonator for the TM_{011} mode with a small measurement sample in the bottom center. (a) Cylindrical resonator loaded with a small sample. (b) Distribution of an electric line of force in TM_{011} resonance.

Therefore, in the present case, since the micro-sample is placed at the center of the bottom surface of the cylindrical cavity, the resonance mode TM_{011}, in which the electric flux distribution density is the maximum, is selected at this portion.

Figure 7.5 shows the measurement model in the present case. In a cylindrical cavity with a height l and a diameter of $2a$, in order to determine the stored energy $\overline{W}$ in the cavity, first, we need to derive each component of the electromagnetic field in the cylindrical coordinates (r, φ, z).

As an example of an electromagnetic-field component in this case, the z component of the electric field and the φ component of the magnetic field can be expressed as follows:

$$\left. \begin{aligned} E_z &= E_{\min} J_0 \left(\rho_{01} \frac{r}{a} \right) \cos \frac{\pi z}{l} \\ H_\varphi &= -E_{\min} j\omega\varepsilon_0 \frac{a}{\rho_{01}} J_0' \left(\rho_{01} \frac{r}{a} \right) \cos \frac{\pi z}{l} \end{aligned} \right\} \quad (0 \le z \le l). \tag{7.45}$$

Here, $J_m(x)$ is a Bessel function and ' denotes its differentiation. In general, ρ_{mn} is the root of $J_m(x)=0$; so in this case, ρ_{01} is the first root of $J_0(x) = 0$. Since a minute sample is placed in the center of the cavity bottom $(0, 0)$ in the TM_{011} mode, the electric field in the sample may be considered as uniform, and its value takes a maximum value E_{max}.

Substituting the H_φ component from Eq. (7.45) into Eq. (7.37), $\overline{W}$ can be represented in the cylindrical coordinate system as follows:

$$\overline{W} = 2\overline{W}_m = \frac{\mu_0}{2} \int_0^a \int_0^l |H_\varphi|^2 2\pi r\, dr\, dz$$

$$= \frac{\pi}{4} \omega_1^2 \varepsilon_0^2 \mu_0 a^4 l \frac{J_1^2(\rho_{01})}{\rho_{01}} |E_{max}|^2. \tag{7.46}$$

(See Appendix 7.A.2.)

Thus, the expressions (7.43) and (7.44) are

$$\frac{\delta f}{f_1} \approx -\varepsilon_0 \chi_{er} \frac{|E_{max}|^2}{4\overline{W}} \Delta V = -\varepsilon_0(\varepsilon_r - 1) \frac{|E_{max}|^2}{4\overline{W}} \Delta V, \tag{7.47}$$

$$\delta\left(\frac{1}{2Q}\right) \approx \varepsilon_0 \chi_{ei} \frac{|E_{max}|^2}{4\overline{W}} \Delta V = \frac{\sigma}{\omega_0} \frac{|E_{max}|^2}{4\overline{W}} \Delta V. \tag{7.48}$$

Substituting the relationship of Eq. (7.46) into Eqs. (7.47) and (7.48),

$$\frac{\delta f}{f_0} \approx -(\varepsilon_{r1} - 1) \frac{1}{4\pi^3} \frac{\rho_{01}^2}{J_1^2(\rho_{01})} \frac{\lambda_0^2 \Delta V}{a^4 l}. \tag{7.49}$$

(See Appendix 7.A.3.)

$$\delta\left(\frac{1}{2Q}\right) \approx \frac{\sigma}{\omega_0 \varepsilon_0} \frac{1}{4\pi^3} \frac{\rho_{01}^2}{J_1^2(\rho_{01})} \frac{\lambda_0^2 \Delta V}{a^4 l}. \tag{7.50}$$

(See Appendix 7.A.4.)

Notice here that, when expressing, generally, ρ_{mn}, this value means the nth root of $J_m = 0$.

Now, $\rho_{01} = 2.4048$ and $J_1(\rho_{01}) = 0.5191$. In Eqs. (7.49) and (7.50), all values except relative permittivity ε_r and conductivity σ are known numbers. Therefore, we can determine these values of real relative permittivity ε_r and conductivity σ by knowing the values of δf and Q in the left-hand side in Eqs. (7.49) and (7.50), respectively.

7.1.2.2 Complex Permittivity Measurement

In the same way when a thin rodlike sample was inserted into the cylindrical cavity resonator, we obtain the material constant in the sample by measuring the change in the value of the resonant frequency δf and the quality factor Q of the cavity resonator. Let us describe here a method of obtaining a complex dielectric constant in the case where a thin rodlike dielectric is placed in the center axis direction of the cavity resonator, as shown in Figure 7.6.

In this case, the TM_{010} mode in the cylindrical cavity resonator is considered.

As shown in Figure 7.6, the electric field is maximized at the central portion in the present cylindrical coordinate system (r, φ, z), and the magnetic fields are distributed so as to surround the central axis z in the case of the TM_{010} mode.

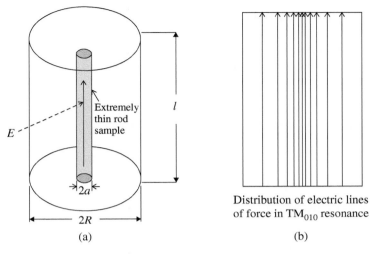

(a)

Distribution of electric lines
of force in TM$_{010}$ resonance

(b)

Figure 7.6 Dielectric constant measurement by the TM$_{010}$ mode using a small rod.
(a) Cylindrical resonator loaded with a small dielectric sample. (b) Distribution of an electric
line of force in TM$_{010}$ resonance.

Therefore, the complex dielectric constant of the sample is determined due to this electromagnetic-field distribution by inserting a narrow dielectric rod in the direction of the cavity center. Because the dielectric radius is extremely small, the Z-axis component of the electric field is almost unchanged.

The cavity resonator inside is assumed to be vacuum, the dielectric constant ε is represented by $\varepsilon = \varepsilon_0$, and E_1 is assumed to be replaced by E_0. Therefore, Eq. (7.35) in the previous section can be used under the assumption of a narrow dielectric rod with the material constants ($\varepsilon = \varepsilon_1$, $\mu = \mu_0$):

$$-\frac{\delta\omega_0}{\omega_0} = \frac{1}{2}\left(\frac{\varepsilon_1}{\varepsilon_0} - 1\right)\frac{\int_{\Delta V} E_1 \cdot E_0^* dV}{\int_V E_0 \cdot E_0^* dV}. \tag{7.51}$$

The electromagnetic-field component of this resonance mode TM$_{010}$ is irrelevant to the φ coordinate in the circumferential direction and to the z coordinate in the rod axis direction.

Hence, the electromagnetic fields in the resonant TM$_{010}$ mode without insertion of a dielectric rod can be represented as follows:

$$E_z = KJ_0(u_{01}r), \tag{7.52}$$

$$H_\varphi = K_1 J_0'(u_{01}r), \tag{7.53}$$

$$(E_r = E_\varphi = H_r = H_z = 0).$$

Here, $u_{01} = \rho_{01}/a$, ρ_{01} is Bessel function's valuable.

Although the constant K can be expressed as $K = \lambda_0 \rho_{01} K_{01} / 2\pi \sqrt{\varepsilon}$, $K_{01} = \sqrt{\varepsilon_s} / \sqrt{\pi l r} J_1(\rho_{01})$, these constants K, K_{01} will disappear in the final expressions, which we are aiming to derive. By assuming that the numerator in Eq. (7.51) is $E_1 \cong E_0$, and substituting Eq. (7.52) into the numerator of Eq. (7.51),

$$\int_{\Delta V} E_1 \cdot E_0^* dV = 2 \int_{\Delta V} |E_0|^2 dV \cong 2K^2 \pi a^2 l. \tag{7.54}$$

In the same way, substituting Eq. (7.52) into the denominator of Eq. (7.51),

$$\int_V E_0 \cdot E_0^* dV = 2 \int_V |E_0|^2 dV = 2K^2 \int_0^L \int_0^{2\pi} \int_0^R J_0^2(u_{01}r) r \, dr \, d\theta \, dz$$
$$= K^2 2\pi L R^2 J_1^2(u_{01}R). \tag{7.55}$$

Accordingly, the following equation can be derived:

$$-\frac{\delta\omega_0}{\omega_0} = \frac{1}{2J_1^2(\rho_{01})} \left(\frac{a}{R}\right)^2 \left(\frac{\varepsilon_1}{\varepsilon_0} - 1\right). \tag{7.56}$$

Here, $\rho_{01} = 2.405$ is obtained as the root of $J_0(\rho_{10}) = 0$ based on the boundary condition of the cylindrical cavity. Accordingly, $J_1(\rho_{01}) = 0.5191$.

Hence, Eq. (7.56) is

$$-\frac{\delta\omega_0}{\omega_0} = 1.855 \left(\frac{a}{R}\right)^2 (\varepsilon_{r1}' - 1 - j\varepsilon_{r1}''). \tag{7.57}$$

As a next step, inserting the sample into the internal cavity of the resonator, the complex resonant angular frequency of the cavity is varied slightly. When this complex resonant angular frequency ω_1 is generally expressed in the same way as in the previous expression in (7.42),

$$\omega_1 = \omega_r + j\omega_i = \omega_r + j(\omega_r/2Q).$$

Using this relationship,

$$\frac{\delta\omega_0}{\omega_0} \cong \frac{\omega_{r1} - \omega_{r0}}{\omega_{r0}} + \frac{j}{2} \left(\frac{1}{Q_{L1}} - \frac{1}{Q_{L0}}\right). \tag{7.58}$$

Here, Q_{L0} and Q_{L1} are the Q values before and after insertion of the sample in the resonant cavity, respectively. The real part and the imaginary part of the complex permittivity are given using Eqs. (7.57) and (7.58), respectively,

$$\frac{\delta\omega_{r0}}{\omega_{r0}} = -\alpha(\varepsilon_{r1}' - 1)\frac{\Delta V}{V}, \tag{7.59}$$

$$\frac{1}{Q_{L1}} - \frac{1}{Q_{L0}} = 2\alpha\varepsilon_{r1}''\frac{\Delta V}{V}. \tag{7.60}$$

Here, $\varepsilon_{r1} = \varepsilon_{r1}' - j\varepsilon_{r1}''$. $\alpha = 1.855$ is the coefficient value of Eq. (7.57), V is the volume of the cavity resonator, ΔV is volume of the sample.

In this way, the resonant frequency changes before and after the sample insertion, and similarly does the Q, so that the complex dielectric constant of the sample can be measured.

7.1.2.3 Complex Permeability Measurement

Let us here investigate a method for determining the complex permeability in the case when the small rod-shaped magnetic material is inserted into a cylindrical cavity resonator, as shown in Figure 7.7. In this case, the TE_{011} mode should be taken as a resonant mode that can produce a magnetic field distribution in the Z-axis direction.

When considering that the radius of the short rod sample is sufficiently small and the variation of the magnetic field in this region is also small, we can apply again Eq. (7.35) from the previous section. We now assume $\varepsilon_{r1} = 1$ in the relative permittivity in Eq. (7.35). Then, substituting the relationship of $\varepsilon_{r1} = 1$ into Eqs. (7.35) and (7.39), the following equation is obtained:

$$-\frac{\delta\omega_0}{\omega_0} = \frac{1}{2}\left(\frac{\mu}{\mu_0} - 1\right)\frac{\int_{\Delta V}\boldsymbol{H}_1\cdot\boldsymbol{H}_0^*dV}{\int_V\boldsymbol{H}_0\cdot\boldsymbol{H}_0^*dV}. \tag{7.61}$$

Among the electromagnetic fields of the TE_{011} mode, the magnetic field required to calculate the above equation is

$$H_z = KJ_0(u_{(01)}r)\sin\left(\frac{\pi}{l}Z\right) \tag{7.62}$$

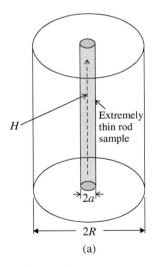

(a)

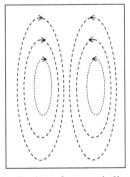

Distribution of magnetic lines of force in TE_{011} resonance

(b)

Figure 7.7 Permeability measurement by the TE_{011} mode using a small rod. (a) Cylindrical resonator loaded with small dielectric sample. (b) Distribution of a magnetic line of force in TE_{011} resonance.

$$H_r = K_1 J_0'(u_{(01)}r) \cos\left(\frac{\pi}{l}Z\right). \tag{7.63}$$

First, the numerators of Eq. (7.61) are assumed so as to approximate H_1 by H_0.

Namely,

$$\int_{\Delta V} H_1 \cdot H_0^* dV = \int_{\Delta V} 2|H_0|^2 dV = \int_{\Delta V} 2(H_z^2 + H_r^2) dV. \tag{7.64}$$

Equation (7.64) is

$$2\int_0^2 \int_0^{2\pi} \int_0^a \left\{ K^2 J_0^2(u_{(01)}r)\sin^2\left(\frac{\pi}{l}Z\right) + K_1^2 J_0'^2(u_{(01)}r)\cos^2(lZ) \right\} r\, dr\, d\theta\, dz$$
$$= K^2 \pi a^2 l. \tag{7.65}$$

Then, in the same way, the denominator of Eq. (7.61) is

$$\int_V H_0 \cdot H_0^* dV = \int_V 2|H_0|^2 dV = \int_V 2(H_z^2 + H_r^2) dV. \tag{7.66}$$

After substituting Eqs. (7.62) and (7.63) into Eq. (7.66) and by performing the integration,

$$2\int_0^L \int_0^{2\pi} \int_0^R \left\{ K^2 J_0^2(u_{(01)}r)\sin^2\left(\frac{\pi}{l}Z\right) + K_1^2 J_0'^2(u_{(01)}r)\cos^2 l\left(\frac{\pi}{l}Z\right) \right\}$$
$$= K^2 \pi R^2 J_0(u_{(01)}R) J_2(u_{(01)}R) \left\{ \frac{J_0(u_{(01)}R)}{J_2(u_{(01)}R)} - \left(\frac{K_1}{K}\right)^2 \right\}. \tag{7.67}$$

Since $u_{(01)}$ is expressed in the form of $u_{(01)} = \rho_{01}/R$, the notations of the Bessel function can be denoted by $J_0(\rho_{01})$ and $J_2(\rho_{01})$. This root of $J_2(\rho_{01}) = 0$ is given by $\rho_{10} = 3.832$ in the present case. Equations (7.65) and (7.67) are substituted into Eq. (7.61), and after making the arrangement of this equation,

$$-\frac{\delta\omega_0}{\omega_0} = \frac{3.094}{\{1 + (\pi R/\rho_{01}l)^2\}}\left(\frac{a}{R}\right)^2 (\mu_{r1}' - 1 - j\mu_{r1}''). \tag{7.68}$$

Further, separating the real part and the imaginary part of Eq. (7.68), the following equation is obtained:

$$\frac{\delta\omega_r}{\omega_{r0}} = -\alpha(\mu_{r1}' - 1)\left(\frac{\Delta V}{V}\right), \tag{7.69}$$

$$\frac{1}{Q_{L1}} - \frac{1}{Q_{L0}} = 2\alpha\mu_{r1}''\left(\frac{\Delta V}{V}\right), \tag{7.70}$$

where $\alpha = 3.094/\{1 + (\pi R/\rho_{01}l)^2\}$.

As is obvious from these discussions, we can determine the electrical material constants, such as permittivity, conductivity, and permeability values using the resonant phenomena in the cavity. In these examples, only an isotropic medium was investigated, but the tensor permeability of an anisotropic material can also be measured even when using a cavity resonator.

7.2 Measurement of EM-Wave Absorption Characteristics

As is well known, the measurement of the characteristics of the EM-wave absorber is usually carried out in an anechoic chamber. However, when the material is in the stage of prototype or where a large structure cannot be constructed, the following method is helpful. Various kinds of methods were proposed for the measurement of the EM-wave absorption characteristics. The main measurement methods are classified into the following types.

(a) The method of using the TEM mode transmission line system,
(b) The waveguide method,
(c) The space standing-wave method,
(d) The space reflected wave method, and the like.

In this section, let us explain these methods which have been used so frequently from the practical viewpoint.

7.2.1 Method of Using TEM Mode Transmission Line

This method is based on measuring a standing wave using a TEM mode, which is equivalent to the plane wave propagation. As is well known, the transmission lines consisting of two conductor systems (at most three conductors), as shown in Figure 7.8, do not have the propagation direction components of the electromagnetic field.

That is, the electric field and the magnetic field are distributed only in the plane perpendicular to the propagation direction. In this case, the electric line of force and magnetic field line of force are orthogonal, and it is well known to be called as the field distributions of the TEM mode. This has the equivalent relationship with a plane wave in free space. Therefore, by loading the EM-wave absorber at the end terminal part of these waveguides, the EM-wave absorber characteristic can be found by measuring the standing

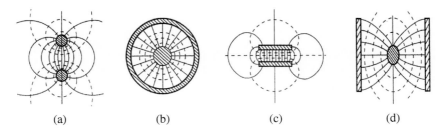

(a)   (b)   (c)   (d)

Figure 7.8 Examples of TEM mode transmission line and waveguide. (a) Parallel two-wire transmission line. (b) Coaxial waveguide. (c) Parallel plate waveguide. (d) Three-conductor type waveguide.

wave. As these representative examples, let us introduce the coaxial waveguide method, the strip line method, and the method of the TEM cell.

7.2.1.1 Coaxial Waveguide Method

Figure 7.9 shows the most basic measurement method which uses the standing-wave meter. This is a method for measuring a standing wave by loading a measurement sample to a coaxial waveguide termination, as shown in Figure 7.9c. In the front part of the sample, the reflected wave and the incident wave are mixed, and the standing wave stands, as shown in Figure 7.10b. Taking the ratio of the maximum value E_{max} of the electric field in the standing wave to the minimum value E_{min}, the standing-wave ratio ρ can be determined. As mentioned previously, the standing-wave ratio is

$$\rho = \frac{E_{\min}}{E_{min}} = \frac{1 + |S|}{1 - |S|}. \tag{2.25}$$

When using a network analyzer, the measurement method of the standing wave, that is, the reflection coefficient measurement, becomes simplified, as shown in Figure 7.9a. The coaxial waveguide method is used, in principle, in the EM wave at low frequency. Therefore, the method is utilized mainly as the measurement method in the microwave bands or lower bands. Also, there is a limitation that it can be measured only in the case of the normal incidence case to the sample plane.

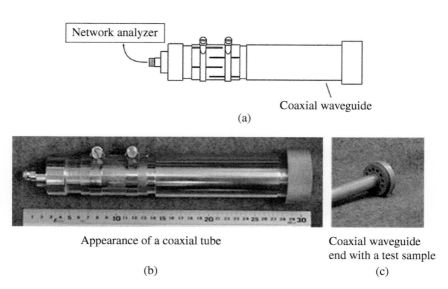

(a)

Appearance of a coaxial tube

(b)

Coaxial waveguide end with a test sample

(c)

Figure 7.9 Reflection coefficient measurement by a network analyzer.

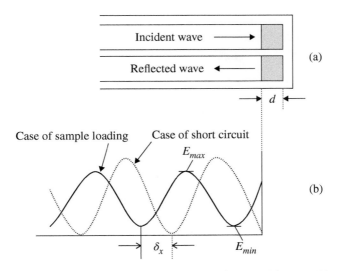

Figure 7.10 Explanation of standing-wave measurement by a coaxial waveguide method.

7.2.1.2 Strip Line Method

Strip line technique is also a method utilizing the TEM wave, where only a normal incident characteristic on the sample can be measured, as shown in Figure 7.11. However, the measurement becomes possible close to the states of the actual EM-absorber size and shape, compared to the coaxial waveguide method. In addition, its structure is also widely used because it is relatively simple. Figure 7.11 shows that the TEM wave, which is guided to the connector portion of the strip line from oscillator, can propagate without reflection due to the taper waveguide being extended gradually, and can reach the

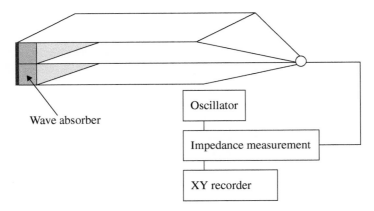

Figure 7.11 Measurement of EM-wave absorption characteristics using a strip line.

measurement material mounted at the strip line termination. Measurement samples are usually loaded in two divided forms, so as to be in close contact with a conductive plate of the waveguide termination. In this type of method, the introduction of time-domain measurement method is preferable to remove the reflected wave generated on each portion.

7.2.1.3 TEM Cell Method

Although both sides of the strip line are open in the lateral directions, alternatively, the TEM cell has a structure that makes the outer conductor of the coaxial waveguide square, and its inner conductor is formed in a flat plate shape. Since the TEM cell is completely shielded, accordingly, without the influence of an external electromagnetic field, it can realize a relatively stable measurement. This type of TEM cell proposal dates back to 1970 [2]. However, on the other hand, the TEM cell becomes a larger structure and becomes more expensive than the strip line. This measurement method is also limited to the case where EM waves are a normal incidence on the EM-wave absorber.

The basic structure of the TEM cell is shown in Figure 7.12. When using the TEM cell to measure the absorber reflection, the matching load is connected to the connector terminal end.

The TEM cells are generally introduced for the use of immunity characteristic measurement of various kinds of electronic devices. When applying the TEM cell to the measurement of the EM-wave absorber, the tapered structure of the end portion of the TEM cell is removed and this terminal end is constructed by a flat conductor plate, as shown in Figure 7.13. That is, this measurement method becomes similar to the coaxial waveguide.

Note here that the electromagnetic-field distribution of the TEM waveform in the TEM cell is limited at a low frequency, depending on its structure size,

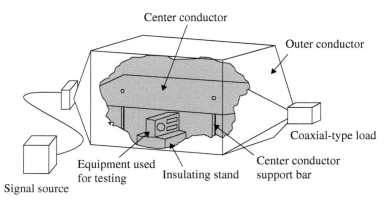

Figure 7.12 Example of a configuration form of a TEM cell.

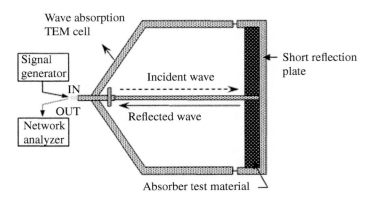

Figure 7.13 Example of configuration when measuring radio wave absorber characteristics of a TEM cell.

and the high-order mode is generated at a high frequency. In order to deal with this problem, GTEM cells have been proposed [3, 4].

7.2.2 Waveguide Method

In the method using the waveguide, the measurement of a small test sample, as well as a relatively large one, such as pyramidal absorbers, is possible. An example of using a large-sized waveguide is described in this section. As shown in Figure 7.14, first of all, a measurement test sample is loaded so as to be in close contact with the waveguide terminal of the short-circuit conductor. And then, this waveguide is equipped with a slot on the waveguide H surface, and the standing-wave measurement probe is moved in this slot to measure the standing waves. When using a large waveguide, the electromagnetic-wave absorption characteristics can be evaluated up to a low frequency of about 70 MHz. But, in this method, because the characteristic impedance is different from the free space impedance value, it is necessary to notice the problem that does not always coincide with the wave absorption characteristics of free space.

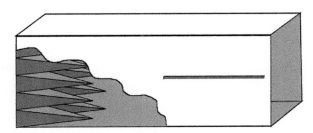

Figure 7.14 Measurement of wave absorption characteristics due to the waveguide.

7.2.3 Space Standing-Wave Method

The space standing-wave method is classified into two types. One is the method of measuring a standing wave by moving a receiving antenna or an electric field probe between a transmitting antenna and an EM-wave absorber. The other is the method of fixing the receiving antenna and moving the EM-wave absorber. First, taking here the former case as an example, the basic idea of measurement is described. As shown in Figure 7.15a, let us consider here the case where the incident EM wave is assumed to be perpendicular to the measurement sample which is placed in the space. Since there usually exist some reflections from the test sample, the reflected wave in the sample front and the incident wave form standing waves. In practice, however, since the electric field radiated from the antenna is attenuated, depending on the distance, the standing wave of uniform

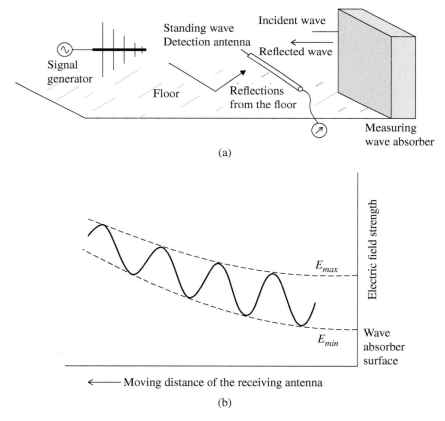

(a)

(b)

Figure 7.15 Spatial standing-wave method. (a) Layout drawing of the space measurement. (b) The standing wave of the actual space.

amplitude cannot often be measured. Therefore, along with the distance from the sample surface, the waveforms decay, as shown in Figure 7.15b [1].

As shown in Figure 7.15, to get the reflection coefficient in this case, it is necessary to draw the envelope of the standing wave, as indicated by the dotted line, and get the upper and lower envelope values E_{max} and E_{min}, respectively. Then, the reflection coefficient can be calculated from the relation of the standing-wave expression in (2.25), using the E_{max} and E_{min} of the absorber surface.

Next, the basic configuration method of measuring the spatial standing wave in practical use is illustrated in Figure 7.16a,b [5, 6]. These are the examples

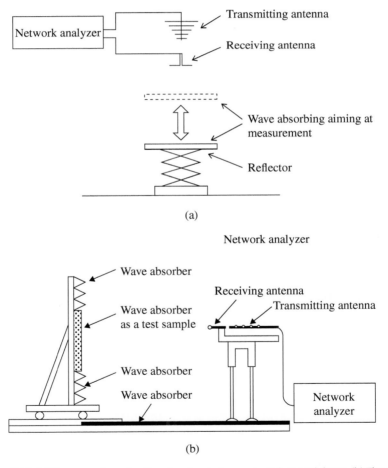

(a)

(b)

Figure 7.16 (a) The case of moving the absorber to be measured up and down. (b) The case of moving the absorber to be measured in the horizontal direction.

of a case where the receiving antennas are fixed and the EM-wave absorbers are moved. It should be noted that the methods mentioned are the measurement methods of the ideal case in which the reflected wave is considered not to exist at the floor, ceiling, and side walls. In practice, cases where we can ideally measure the reflection coefficient are almost non-existent, but the analytical processing methods described in Section 3.4.1 of Chapter 3 have to always be kept in our mind. Here, from the practical point of view, the spatial measurement methods of introducing network analyzers equipped with a time domain measurement system were introduced.

7.A Appendix

7.A.1 Appendix to Section 7.1.2 (1)

Derivation of Eq. (7.35)

$$\nabla \times E_0 = -j\omega_0 \mu H_0 \tag{7.A.1}$$

$$\nabla \times H_0 = j\omega_0 \varepsilon E_0 \tag{7.A.2}$$

$$\nabla \times E = -j\omega_1 (\mu + \Delta\mu) H \tag{7.A.3}$$

$$\nabla \times H = j\omega_1 (\varepsilon + \Delta\varepsilon) E. \tag{7.A.4}$$

From Maxwell's Eqs. (7.A.1)–(7.A.4), the following relations are obtained:

$$E_0^* \cdot (\nabla \times H) + H \cdot (\nabla \times E_0)^* = j\omega_1 (\varepsilon + \Delta\varepsilon) E \cdot E_0^* - j\omega_0 \mu H_0^* \cdot H, \tag{7.A.5}$$

$$E \cdot (\nabla \times H_0)^* + H_0^* \cdot (\nabla \times E)^* = j\omega\varepsilon E_0^* \cdot E - j\omega_1 (\mu + \Delta\mu) H_0^* \cdot H. \tag{7.A.6}$$

Using these relations in expressions (7.A.5) and (7.A.6), we apply the volume integral to the equation after subtracting (7.A.5) and (7.A.6).

As a result, the integral on the left side becomes 0 from the relationship between $E//n$ and $E_0^*//n$.

On the other hand, the right side of this volume integral,

$$\int_V (\omega_1 - \omega_0)\varepsilon E \cdot E_0^* dV + \int_V (\omega_1 - \omega_0)\mu H \cdot H_0^* dV$$

$$+ \omega_1 \int_{\Delta V} (\Delta\varepsilon E \cdot E_0^* + \Delta\mu H \cdot H_0^*) dV = 0. \tag{7.A.7}$$

From the expression (7.A.7), the relationship of Eq. (7.35) can be derived.

7.A.2 Appendix to Section 7.1.2 (2)

Derivation of Eq. (7.46).
Considering the mode of TM_{011} in the cylindrical cavity resonator, the magnetic field in Eq. (7.46) is

$$H_\varphi = -E_{max}j\omega\varepsilon_0\frac{a}{\rho_{01}}J_0'\left(\rho_{01}\frac{r}{a}\right)\cos\frac{\pi z}{l}. \tag{7.A.8}$$

Substituting this equation into Eq. (7.46) and transforming it,

$$\overline{W} = 2\overline{W}_m = \frac{\mu}{2}\int_V \mu|H|^2 dV$$

$$= \frac{\mu_0}{2}\int_0^a \int_0^{2\pi}\int_0^l |H_\varphi|^2 J\, dr\, d\varphi\, dz$$

$$= \mu_0 \int_0^a \int_0^l |H_\varphi|^2 r\, dr\, d\varphi\, dz. \tag{7.A.9}$$

Here, we can denote the Jacobian matrix J as $J = r$ and μ can be expressed like μ_0.

Substituting the relationship of $\int_0^l \cos^2\left(\frac{\pi z}{l}\right) dz = \frac{l}{2}$ into Eq. (7.A.9),

$$\overline{W} = \frac{1}{2}|E_{max}|^2\mu_0\varepsilon_0^2\omega^2 l\pi\left(\frac{a}{\rho_{01}}\right)^2 \int_0^a \left[J_0'\left(\rho_{01}\frac{r}{a}\right)\right]^2 r\, dr. \tag{7.A.10}$$

Here, when the variable of the Bessel function is replaced by $X = \frac{\rho_{01}}{a}r$, the integral term of (7.A.10) becomes

$$\int_0^a [J_0'(\rho_{01})]^2 r\, dr = \left(\frac{a}{\rho_{01}}\right)^2 \int_0^{\rho_{01}} [J_0'(X)]^2 X\, dX. \tag{7.A.11}$$

Further, the integral formula of the Bessel function expressed by the following equation is applied to Eq. (7.A.11):

$$\int_0^{\rho_{01}} [J_0'(X)]^2 X\, dX = \int_0^{\rho_{01}} J_1^2(X) dX$$

$$= \frac{\rho_{01}}{2}J_1^2(\rho_{01}). \tag{7.A.12}$$

As a result, Eq. (7.46) can be derived:

$$\overline{W} = \frac{\pi}{4}\omega_1^2\varepsilon_0^2\mu_0 a^4 l\frac{J_1^2(\rho_{01})}{\rho_{01}}|E_{max}|^2.$$

7.A.3 Appendix to Section 7.1.2 (3)

Derivation of Eq. (7.49).

From Eq. (7.46)

$$\frac{|E_{max}|^2}{4\overline{W}} = \frac{\rho_{01}}{\pi\omega_0^2\varepsilon_0^2\mu_0 a^4 l J_1^2(\rho_{01})}$$

$$= \frac{\lambda_0^2\rho_{01}}{4\pi^3\varepsilon_0 l a^4 J_1^2(\rho_{01})}. \tag{7.A.13}$$

Here, the following relationships are applied to Eq. (7.A.13):

$$\left(\sqrt{\varepsilon_0\mu_0}\right)^2\varepsilon_0 = \frac{1}{c^2}\varepsilon_0$$

$$\omega_0^2 = 4\pi^2\frac{c^2}{\lambda_0^2}.$$

When Eq. (7.A.13) is substituted into Eq. (7.47), Eq.(7.49) is obtained.

7.A.4 Appendix to Section 7.1.2 (4)

Derivation of Eq. (7.50)

First, the complex permittivity is expressed by the following equation:

$$\dot{\varepsilon}_0\varepsilon_r = \varepsilon - j\left(\sigma_d/\omega\right). \tag{7.A.14}$$

Using this relationship, the relative polarizability is expressed by the following equation:

$$\varepsilon_r - 1 = \varepsilon_r - 1 - j(\sigma_d/\varepsilon_0\omega)$$

$$= \chi_{er} - j\chi_{ei}. \tag{7.A.15}$$

As a result, the imaginary part of the relative polarizability is given by the following equation:

$$\therefore \chi_{ei} = \sigma_d/\varepsilon_0\omega. \tag{7.A.16}$$

Substituting χ_{ei} in Eq. (7.A.16) into χ_{ei} into Eq. (7.48) yields Eq. (7.50).

References

1 Shimizu, Y. (Editorial Committee Chairman) (1999). *Electromagnetic Waves Absorption and Shielding*, 123. Tokyo: Nikkei Gijyutu Tosho.
2 Crawford, M.L. (1974). Generation of standard EM fields using TEM transmission cells. *IEEE Trans. Electromagn. Compa.* EMC-16 (4): 189–195.
3 Konigstein, D. and Hansen, D. (1987). A new family of TEM cells with enlarged bandwidth optimized working volume. In: *Proceedings 7th, International Zurich Symposium and Technology Exhibition of EMC*, 127–132.

4 Hansen, D., Wilson, P., Konigstein, D., and Schaer, H. (1989). A broadband alternative TEM test chamber based on TEM-cell anechoic -chamber hybrid concept. In: *International Symposium on EMC (EMC' 89/Nagoya)*, vol. 1, 133–137.

5 Tada, M. and Morozumi, M. (1995). Construction of TV-ghost suppression ferrite absorbing panels. In: *Summaries of Technical Papers of Annual Meeting Architectural Institute of Japan*, 1067.

6 Nakagawa, H., Kobayashi, M., and Kasashima, Y. (1995). TV radio wave reflection characteristics on a curtain wall consist of synthetic fiber reinforced concrete. In: *Summaries of Technical Papers of Annual Meeting Architectural Institute of Japan*, 159–160.

8

Configuration Examples of the EM-wave Absorber

In recent years, along with increase in various kinds of demands on communication systems, the communication environment has become more complicated. As a result, the introduction of electromagnetic (EM)-wave absorbers is required in each technical field, especially from the viewpoint of preventing interference and mutual interaction. Regarding the pyramid-type absorber, it is widely used and popularized in anechoic chambers; these technologies have reached a high completion degree. For this reason, in this chapter, EM-wave absorber characteristics are addressed, such as those of the quarter-wavelength absorber, single-layer absorber, and two-layer absorber, while touching upon the implementation principle.

In Section 8.1, a quarter-wavelength absorber actually being put into practical use is introduced along with the theory. Section 8.2 describes the EM-wave absorption characteristics of a single-layer absorber using sintered ferrite and rubber ferrite, together with detailed data.

As a special example, the characteristics of a two-layer absorber developed for a compact anechoic chamber is described in Section 8.3. In Section 8.4, EM-wave absorbers as building materials developed for countermeasures against TV ghosts caused by multiple reflected waves from building walls are concretely explained. Finally, in Section 8.5 is introduced a low-reflection EM-wave absorber that was developed to prevent EM-wave reflections inside buildings from obstructing the operation of handy phone system (PHS) and medium-speed wireless local-area network (LAN).

8.1 Quarter-wave-Type Absorber

A quarter-wave-type absorber is considered a suitable example for understanding the basic principle of the EM-wave absorber. This EM-wave absorber consists of a resistive film, a spacer, and a terminating conductor plate in order

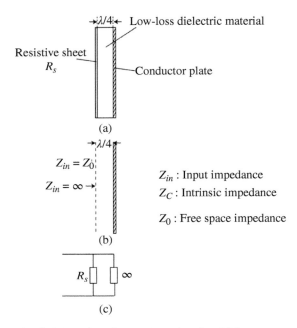

Figure 8.1 Principle of 1/4-wavelength-type wave absorber. (a) Quarter-wavelength type absorber. (b) Explanation of input impedance. (c) The principle of a quarter-wavelength type absorber from the viewpoint of equivalent transmission line.

from the EM-wave incident surface, as shown in Figure 8.1. The absorption principle in this EM-wave absorber can be easily understood by considering the transmission line circuit equivalent to this absorber.

In the transmission line theory, the input impedance Z_{in} when looking at the direction of the conductive plate at a point away from the conductive plate by $\lambda/4$ becomes infinite, as shown in Figure 8.1b. Then, if a thin resistive film having R_s value is placed at a distance $\lambda/4$ from the terminating conductor plate, the relationship in this case can be expressed by the parallel circuit with the R_s and the input impedance $Z_{in}(=\infty)$ in the transmission line, as shown in Figure 8.1b.

In other words, since the synthetic impedance is represented with an infinite impedance $Z_{in} = \infty$ and R_s, the input impedance Z_{in} that looks at the conductive terminal plate from the resistive film is represented by R_s only. Accordingly, when the input impedance Z_{in} is determined so as to take the free space value $Z_{in}(= Z_0 = R_s = 120\pi(\Omega))$, it becomes possible to perform a matching.

In the present case, the reflection coefficient S is represented (refer Eq. (2.21)).

$$S = \frac{Z_{in} - Z_0}{Z_{in} + Z_0} \tag{8.1}$$

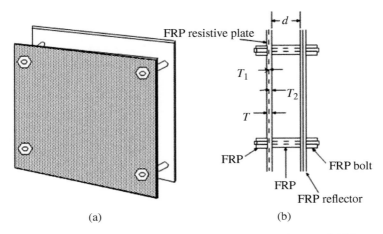

Figure 8.2 Example of a wave absorber made of 1/4 FRP structural material. (a) Front view of 1/4 wavelength type wave absorber. (b) The side view of the absorbent body.

Therefore, the reflection coefficient becomes 0. As a result, it is found that a $\lambda/4$-type EM-wave absorber can be constructed using the sheet resistance value of the resistive film with $120\pi(\Omega)(\approx 377(\Omega))$. Moreover, in practice, since the spacers have certain dielectric values, their distance $\lambda/4$ from the terminating conductor plate should be determined as $\lambda/4 = \lambda_0/4\sqrt{\varepsilon_s}$. ε_s represents the relative permittivity in-between the resistive sheet and the conductive plate.

As an example of a $\lambda/4$-type absorber, the EM-wave absorber constructed by the resistive cloth composed of conductive fibers is shown in Figure 8.2 [1].

In this structure, the resistance cloth, woven with conductive fiber, is sandwiched between two plates by FRP (fiberglass reinforced plastics) plates, which are made of an unsaturated polyester resin reinforced with glass fiber. As this reflection plate, an FRP plate is used by laminating and pressing a brass mesh.

Assuming that the absorbing material thickness T_1 is a constant value and changing the total absorbing material thickness T, the EM-wave absorption characteristics are presented in Figure 8.3. That is, taking the thickness $T_1 = 1.5$ mm as a constant value and changing the total thickness of absorber T as the values of 3.2, 3.7, and 4.2 mm, the relationship between the absorbing center frequency and the thickness d is shown in Figure 8.3. When the absorbing frequency was fixed, as can be seen from the characteristic in Figure 8.3, the thickness of the total absorber d can be reduced in accordance with the increases in the absorbing material thickness T. This characteristic suggests how to produce a thin absorber. For this absorber, a wind pressure test was also carried out; even in an equivalent wind speed 40 m/s to natural wind, the deflection of the structure is as small as 2 mm. For this reason, they are used in various outdoor facilities.

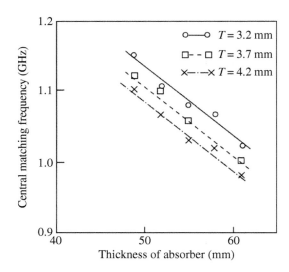

Figure 8.3 Relationship between center frequency and thickness in a $\lambda/4$-type structural absorber.

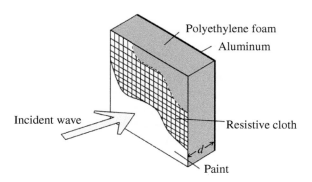

Figure 8.4 A $\lambda/4$-type absorber using a resistive cloth.

Next, as another example of the $\lambda/4$-type absorber, the characteristics of a prototype made for a ship radar using a resistive fabric cloth woven with conductive fibers as shown in Figure 8.4 are described [2]. Figure 8.5 shows the characteristics of the normal incidence against the thickness of the absorber prototyped for a ship radar (9.4 GHz).

When paying attention to the case of a marine radar at a frequency of 9.4 GHz, this absorber meets the reflection coefficient value of -20 dB in the thicknesses A, B, and C of the specimens. Also, as for the characteristics at oblique incidence for the TM wave, the reflection coefficient is kept

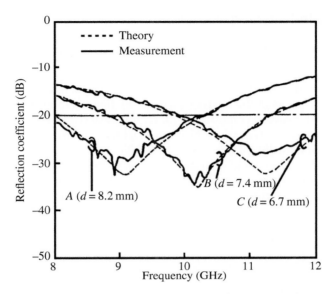

Figure 8.5 Frequency characteristics of the reflection coefficient due to changes in the thickness of the EM-wave absorber.

below −20 dB at the incident wave angles below 30°, as shown in Figure 8.6. This absorber was further improved, and a laminated resistive cloth using chlorosulfonate polyethylene rubber, which has excellent strength and weather resistance, was developed [2].

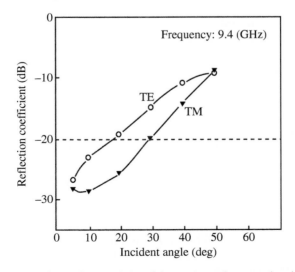

Figure 8.6 Oblique incidence characteristics of the marine radar wave absorber.

8.2 Single-Layer-Type Absorber

8.2.1 Ferrite Absorber

A typical example of a single-layer-type EM-wave absorber is a ferrite absorber.

For ferrite EM-wave absorbers, researches were conducted starting from various angles, such as in the analysis method [3], investigations of the physical properties [4], material composition [5, 6], broadband absorber [7, 8], etc. In addition, since ferrite is relatively heavy in weight, research has proceeded from the viewpoint of making a thin absorber [9, 10].

In general, as a single-layer-type EM-wave absorber, the type using wave-absorbing materials represented by a carbon-based material and the type using a ferrite-type magnetic EM-wave-absorbing materials are mainly used. In the case of a single-layer-type absorber in transverse electric (TE) and transverse magnetic (TM) waves for normal incidence, the configuration dimensions and material constants must be determined so that the reflection coefficient in Eq. (8.2) is 0. That is, this can be achieved by adjusting the relative permittivity ε_{r2}, relative permeability μ_{r2}, and the thickness d of the absorber. Again, showing Eq. (8.2),

$$S = \frac{\sqrt{\frac{\mu_{r2}}{\varepsilon_{r2}}} \tanh\left(j\frac{2\pi}{\lambda}\sqrt{\varepsilon_{r2}\mu_{r2}}d\right) - 1}{\sqrt{\frac{\mu_{r2}}{\varepsilon_{r2}}} \tanh\left(j\frac{2\pi}{\lambda}\sqrt{\varepsilon_{r2}\mu_{r2}} - 1\right) + 1}. \tag{8.2}$$

In the case of using a dielectric absorber material mainly composed of a carbon material, if the relative permeability is set to 1.0 and the reflection coefficient S in Eq. (8.2) is set to 0, the thickness d can be determined using a known dielectric constant.

In this connection, by obtaining the value that makes $S = 0$ for both parameters ε_2 and d/λ_0, the relationships between the relative permittivity and the thickness can be found. However, in magnetic materials like ferrite, both μ_{r2} and ε_{r2} have a certain value, respectively, and change with frequency, so it is needed to determine the thickness d by obtaining these precise values. As an example of a single-layer-type absorber, let us introduce here the examples of EM-wave absorbers with sintered ferrite and rubber ferrite, respectively.

First, when the sintered ferrite is compared with rubber ferrite, the relative magnetic permeability of the sintered ferrite takes a large value and is often used for an EM-wave absorber from approximately 30 MHz to 2.0 GHz. The ferrite thickness for use as an absorber is about 4–10 mm. The single-layer-type EM-wave absorber using sintered ferrite has a figure of merit (refer to Chapter 1) and it is usually about 80% and 150%.

On the other hand, the rubber ferrite is a material in which sintered ferrite powder or a material called a by-product ferrite is blended in a rubber that is flexible and easy to process.

Table 8.1 Characteristics of sintered ferrite absorber.

Types of ferrite	Thickness d (mm)	Frequency below −20 dB (MHz)	Absorbing central frequency f_0 (MHz)	$\Delta f / f_0$ (%)
Ni–Zn series	7.0	80–360	190	147
Mn–Mg–Zn series	7.5	250–800	410	134
Ni–Zn	4.0	600–1270	860	78

Table 8.2 Characteristic example of rubber ferrite absorber (Ni–Zn system, from TDK Corporation data).

Volume mixing ratio of ferrite (%)	Thickness d (mm)	Less than −20 dB in the frequency range (GHz)	Absorbing central frequency f_0	$\Delta f / f_0$ (%)
45	5.8	2.7–3.75	3.2	33
36	5.8	3.4–4.5	4	28
33	5.0	4.6–6.0	5.3	26
31	4.8	5.5–7.4	6.35	30
35	4.2	6.4–8.6	7.75	28
45	3.4	9.0–10.2	9.7	12

From the viewpoint of mixing ferrite powders in the rubber, the relative permeability cannot take a large value compared with the sintered ferrite, so that the EM-wave absorption band is above 1 GHz. Also, the figure of merit of typical rubber ferrite is about 30%. Both characteristics are shown in Tables 8.1 and 8.2.

8.3 Two-Layered Absorber

In the absorber of the two-layer structure, the material characteristics of the EM-wave absorber and the thickness of each layer are changed in such a way that the degree of freedom of characteristic control can be expanded. Moreover, a two-layer absorber can possess a feature that can be configured by an EM-wave absorber with a wideband frequency of absorbing characteristics than those of a single-layer structure, and also have a feature that the matching frequency bands can be easily changed.

Figure 8.7 shows a two-layer structure absorber that was developed for a small anechoic chamber [11]. Ferrite is arranged on the side of the EM-wave incident surface, and a dielectric board of calcium silicate system is mounted

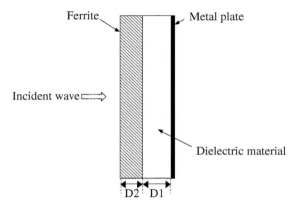

Figure 8.7 Structure of the two-layer-type ferrite absorber.

on its back, as shown in Figure 8.7. This EM-wave absorption mechanism can be easily understood on the Smith chart.

In this configuration, when increasing the thickness of the dielectric material, the input normalized impedance to the absorber, especially in the higher frequency region, can be shifted from the capacitive region on the Smith chart to the inductive region. Meanwhile, when selecting the ferrite material having characteristics that the normalized impedance value is close to 1.0 and the ferrite resistance value is relatively small, the normalized impedance for the absorber has the tendency of circulating around 1.0 while moving toward the inductive region on the Smith chart. As a result, this EM-wave absorber exhibits relatively broadband characteristics. Focusing on these

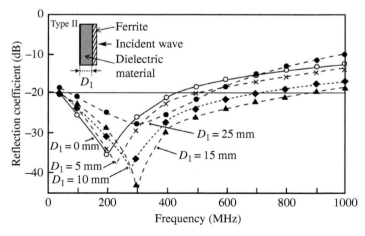

Figure 8.8 Frequency characteristics of the reflection coefficient of a two-layer type absorber.

characteristics, the frequency characteristic of the reflection coefficient of the two-layer EM wave absorber when trying to optimize the thickness of the dielectric material D_1 is shown in Figure 8.8. Figure 8.8 shows the case of adopting a dielectric material having a relative dielectric constant $\varepsilon_r = 3.0$. When changing its dielectric material thickness from 10 to 15 mm, a broadband of about 50–800 MHz in the reflection coefficient under −20 dB or less can be realized. Further, below the reflection coefficient −15 dB, the wideband characteristics of 0.02–1.1 GHz can be achieved, and these characteristics satisfy the criteria of CISPR in the EM-wave absorber.

8.4 Applications as Building Material

The EM-wave absorbers of a new type have been proposed mainly as applications to prevent TV ghost images, indoor wireless LANs, mobile communications, etc.

8.4.1 TV Ghost Prevention Measures

In this section, let us outline an example of a concrete countermeasure for a TV ghost on a building wall surface. Let us first consider the TV ghost wave problem around 100 MHz in the very high frequency (VHF) band [12, 13]. This is an example of the Tokyo Metropolitan Government building in Japan, which has ferrite plate materials buried into the granite wall surface layer. In this kind of absorber, there are difficult problems in that it is necessary to harmonize the ideas of the building designers with the EM-wave absorber engineers. Figure 8.9

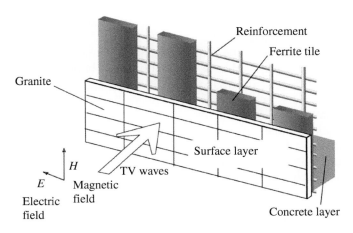

Figure 8.9 Absorber consisting of granite, ferrite layer, and concrete layer [12].

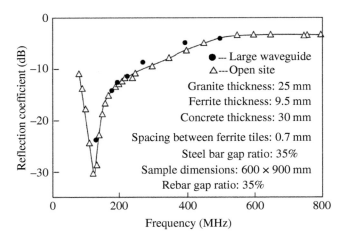

Figure 8.10 Characteristics due to the prototype absorber panel.

shows the absorber layer part consisting of three layers of granite, ferrite, and concrete. In this case, when the thickness of the granite becomes 15 mm or more, a problem arises that the EM-wave absorption characteristics at high frequencies of VHF band deteriorate.

However, in practice, from the viewpoint of the building strength, the granite thickness needs to be 25 mm or more. To satisfy this condition, a means for providing slits between the ferrite tiles is taken, as shown in Figure 8.9. By this means, changing the magnetic resistance of the ferrite layer, the complex relative magnetic permeability is equivalently adjusted. That is, while changing the thickness of the granite, concrete, and ferrite, and optimizing the slit dimensions, the desirable absorption characteristics in the VHF television frequency band can be realized as shown in Figure 8.10. In this case, the optimal slit length between the ferrite tiles is 0.7 mm.

Further, the thickness of the concrete, ferrite, and granite in this case is 30, 9.5, 25 mm, respectively, and have a gap ratio of 35% between the rebars. The reflection coefficient at 100 MHz is −18 dB or less and it is −13 dB below at 200 MHz, which are recognized as sufficient characteristics from the practical use viewpoint.

Next, let us introduce another example of achieving TV ghost measures by means of the so-called curtain wall [14]. This method is based on the idea that the reflected wave causing a TV ghost failure comes only from the metal material portions, such as reinforcing bars in the building wall. Accordingly, a ferrite absorber is loaded only in the portion generating a reflected wave, and the television waves can be transmitted to the indoor side from the other parts with few reflections, such as windows. It is also a proposal focusing on the property

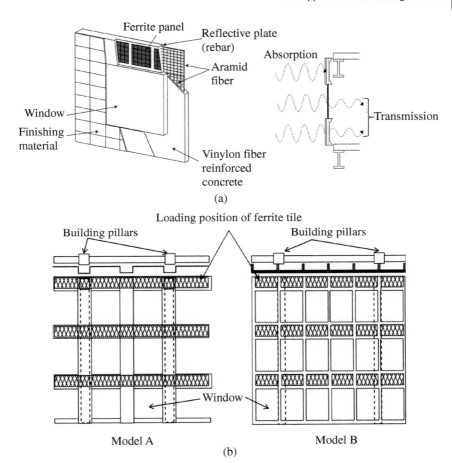

Figure 8.11 Curtain-wall-type absorber configuration. (a) Structural concept of EM-wave transmission-type curtain wall. (b) An example of a ferrite mounting method according to the structural form of a building.

that a concrete made of synthetic fibers can transmit the television EM waves through it. The generic name of this absorber, including the absorbing nature, is referred to as the EM-wave transmission-type curtain wall. This configuration is shown in Figure 8.11. In order to minimize the use of metallic material, aramid fiber reinforcement in the shape of a rod or mesh is used in the vinylon fiber reinforced concrete (VFRC). The real part of the complex relative permittivity of this VFRC is usually about 4–5.

Since the wall of the VFRC curtain is thin and its thickness is between 50 and 70 mm, good transmission characteristics are established for using the TV EM wave. Based on these construction concepts, the curtain wall absorber can be

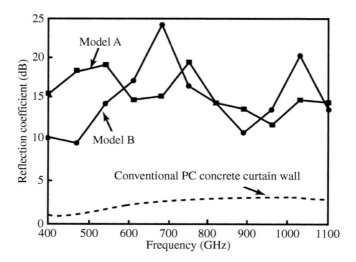

Figure 8.12 Reflection characteristics of the EM-wave transmission-type curtain wall.

realized with the functions by which EM waves arriving at the rebar portion can be absorbed by the ferrite and the other portion of the curtain wall can be transmitted through the EM wave.

Figure 8.12 shows the reflection characteristics measured in the frequency range from 400 to 1100 MHz in the wall surface model, which is shown in Figure 8.11b by the spatial standing-wave method. A reflection coefficient of −10 dB or more has been obtained in this frequency range [15]. In this configuration example of the EM-wave absorber, since the ferrite tile use area is only about half of that of the conventional one, this construction method has a feature that the use of expensive ferrite material can be suppressed.

8.4.2 Ferrite Core-Embedded PC Board

The precast reinforced concrete (PC) board used for the outer building walls is generally reinforced with double reinforcing bars inside the wall. By mounting the ferrite core on one of the reinforcements, a method of reducing the television EM-wave reflection from the building wall has been proposed [15]. This fundamental configuration is shown in Figure 8.13.

As for the mounting method of the ferrite core, each case as in Figure 8.14 has been studied.

Estimating the reflection coefficient based on the theoretical calculation, it becomes possible to find a method to construct a PC board using as few ferrite cores as possible. After estimating the reflection loss from the theoretical calculation, the measurement of absorbing characteristics is conducted on the basis of their data. Figure 8.15 shows the result of measuring the reflection loss

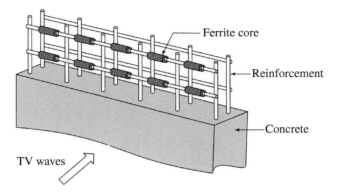

Figure 8.13 Schematic diagram of the PC board of the mounting ferrite core.

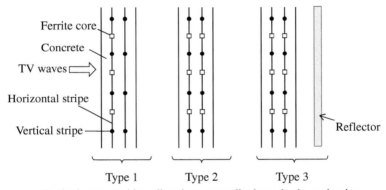

Ferrite is arranged in a direction perpendicular to horizontal stripe.

Figure 8.14 Ferrite core arrangement.

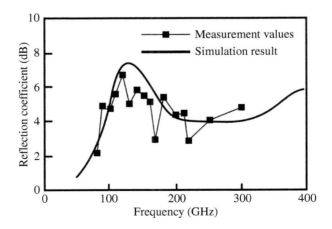

Figure 8.15 EM-wave reflection coefficient of a specimen with ferrite cores.

by the spatial standing wave method by selecting Type 1 with fewer numbers of ferrite cores. Compared with the 100 MHz band, which is required as the television frequency, it is found that the ferrite core–attached specimen has a reflection loss of about 7.5 dB.

8.5 Low-Reflective Shield Building Materials

Recently, reflected waves occurring from various kinds of electronic devices inside building rooms were found to cause problems for the operation of PHS and wireless LAN. As the countermeasure to these EM-wave interference problems, low-reflective shield materials have been developed. Let us introduce here this low-reflective shield material from the viewpoint of understanding this type of EM-wave absorber. As has been described so far, in general, EM-wave absorbers are designed mainly on the basis of a reflection coefficient of less than −20 dB. However, the EM-wave absorber as the building material mentioned herein indicates that the required characteristics are −6 dB or less (approximately).

Let us now introduce this kind of absorber from the electromagnetic compatibility (EMC) standpoint, since this absorber can be evaluated as a valid fault tolerance to the delay spread caused by a personal computer. This absorber material is called a building material to depress the multipath (BMDM). The BMDM configuration is shown in Figure 8.16. Each part of the absorber material is made of a resin containing a ferrite powder, a resin containing an iron powder, and a gypsum board composed of a fire-resistant material; these materials are mounted on the surface of aluminum material termination.

The thickness of each part of the absorber is depicted in Figure 8.16. The theoretical value of the reflection coefficient in this case is shown in

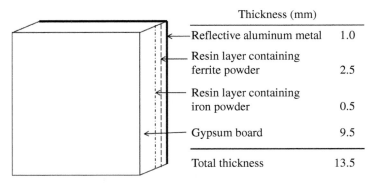

	Thickness (mm)
Reflective aluminum metal	1.0
Resin layer containing ferrite powder	2.5
Resin layer containing iron powder	0.5
Gypsum board	9.5
Total thickness	13.5

Figure 8.16 Construction of a low-reflection-type shielding material.

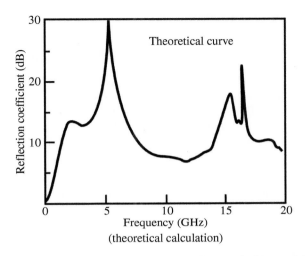

Figure 8.17 Wave absorption characteristics of a low-reflective shield material.

Figure 8.17. It can be confirmed to be more than −6 dB in the frequency band 2.45–19 GHz. Further, as another configuration example, a thinner BMDM has been proposed.

In this absorber material structure, Mn–Zn ferrite powder and carbon fiber are mixed with each gypsum board, respectively, and these gypsum boards are laminated and placed on a terminal end aluminum foil. The characteristics and dimensions of these BMDM and the new BMDM as building materials are shown in Table 8.3. Also, the reflection coefficient characteristics as the building materials at the frequency 1.9, 2.45, and 19 GHz, for each of the TE and TM waves at the time of oblique incidence, are shown in Table 8.4.

The technologies associated with these absorbers mentioned herein, however, will be achieved using the conventional material discussed previously in Chapter 6. In Chapters 9 and 10, the absorber being available at a high frequency leading to the terahertz band will be introduced.

Table 8.3 Characteristics and dimensions of BMDM and the new BMDM.

	Weight (kg/m²)	Thickness (mm)	Thermal resistivity (m²·K/W)	Bending strength (N/cm²)	Compressive strength (N/cm²)	Fireproof ability
BMDM	18	13.5	—	—	—	Semi-incombustible
New BMDM	9	9.5	0.045	1000	451	Incombustible
Gypsum	7	9.5	>0.043	~800	>400	Incombustible

Table 8.4 Reflection loss (dB) of BMDM walls.

Frequency (GHz)	Absorptive material	Incident angle and mode					
		TM			TE		
		30°	45°	60°	30°	45°	60°
1.9	BMDM	14.3	12.0	12.9	9.6	8.7	4.8
	New BMDM	8.3	9.4	12.0	6.8	6.0	5.7
2.45	BMDM	13.5	14.0	15.1	10.6	7.3	4.9
	New BMDM	7.4	10.2	8.8	4.8	3.2	4.0
19	BMDM	13.0	14.5	10.7	10.5	13.3	13.2
	New BMDM	6.7	10.6	12.0	3.5	3.1	2.4

References

1 Hashimoto, Y., Ichihara, K., Ishino, K., and Shimizu, Y. (1989). Practical designing of $\lambda/4$ type microwave absorber with resistive cloth woven by conductive and polyester fiber. *IEICE Trans.* J72-B-II (9): 483–491.

2 Hasimoto, Y., Ichihara, K., Ishino, K., and Shimizu, Y. (1988). $\lambda/4$ type microwave absorber with resistive cloth woven by conductive and polyester fibers. *IEICE Trans.* E76-C (10): 1657–1663.

3 dawson, L., Clegg, J., Porter, S.J. et al. (2002). The use of genetic algorithm to maximize the performance of a partially lined screened room. *IEEE Trans. Electromagn. Compat.* 44 (1): 233–242.

4 Harmuth, H.F. (1985). Use of ferrite for absorption of electromagnetic wave. *IEEE Trans. Electromagn. Compat.* 27 (2): 100–102.

5 Dishovski, N., Petkov, A., Nedkov, I., and Razkazov, I. (1994). Hexaferrite contribution to microwave absorber characteristics. *IEEE Trans. Magn.* 30 (2): 969.

6 Miyata, Y. and Matsumot, M. (1995). Two-layer wave absorber composed of soft-magnetic and ferroelectric substance. *IEEE Trans. Magn.* 33 (5): 3427–3429.

7 Sugimoto, S., Kondo, S., Okayama, K. et al. (1999). M-type ferrite composite as a microwave absorber with wide bandwidth in GHz range. *IEEE Trans. Magn.* 35 (5): 3154–3156.

8 Park, M.J. and Kim, S.S. (1999). Control of complex permeability and permittivity by air cavity in ferrite-rubber composite sheets and their wide-band absorbing characteristics. *IEEE Trans. Magn.* 35 (5): 3184–3183.

9 Musal, H.M. Jr. and Hahn, H.T. (1989). Thin-layer electromagnetic absorber design. *IEEE Trans. Magn.* 25 (5): 3851–3853.

10 Kotsuka Y., and Yamazaki H., Fundamental investigation on a weakly magnetized ferrite absorber, *IEEE Trans Electromagn, Compat.*, pp. 116–124, vol.42, no.2, May 2000.

11 Hashimoto, Y., Ishino, K., and Shimizu, Y. (1990). Practical design of compact anechoic chamber. *IEICE Trans.* J73-B-II (8): 421–431.

12 Hashimoto, Y., Ishino, K., and Shimizu, Y. (1989). Practical investigation on ferrite absorbing panel for TV ghost suppression. EMC International Symposium (Ngoya in Japan), September, 1989.

13 Vandrer Vorst, A., Rosen, A., and Kotsuka, Y. (2006). *RF/Microwave Interaction with Biological Tissue*. Wiley Interscience.

14 Nakagawa, H., Kobayashi, M., and Kasashima, Y. (1995). TV radio wave reflection characteristics on a curtain wall consist of synthetic fiber reinforced concrete. In: *Summaries of Technical Papers of Annual Meeting*, 159–160. Architectural Institute of Japan.

15 Miyazaki, H., Yoshino, R., Sagawa, Y., and Miyake, S. (1995). Method of preventing reflection of TV-waves by concrete curtain walls (no. 2). In: *Summaries of Technical Papers of Annual Meeting*, 1063–1064. Architectural Institute of Japan.

9

Absorber Characteristic Control by Equivalent Transformation Method of Material Constants

As mentioned in the previous chapters, demands for various EM-wave absorbers are rapidly increasing along with the trends toward complicated electromagnetic environments and the development of higher frequency communication equipment. Generally, it has been considered highly difficult to produce new EM-wave absorbers having a desired absorbing frequency and an absorbing characteristic. One reason for this is that realizing an optimum material constant satisfying the required characteristics becomes difficult when manufacturing an EM-wave absorber. That is, in the conventional method, it often becomes difficult to freely control parameters such as the mixing ratio of the EM-wave absorber's raw materials, the firing temperature, the pressure, and the material compositions to realize the appropriate absorber. Furthermore, the problems of accurately measuring material constants and maintaining the accuracy of component sizes become important to the EM-wave absorber in high-frequency use, particularly in the millimeter or terahertz band.

In this chapter, in order to break through these kinds of obstacles, methods are introduced for realizing a new EM-wave absorber having the desired absorbing or frequency characteristics. To do this, design methods are introduced for employing the same material or a conventional material without manufacturing new materials. This way of thinking is called the equivalent transformation method of material constants (ETMMC) [1–3]. Section 9.1 concretely introduces this basic idea and the means of the ETMMC. In Section 9.2, the various methods of realizing absorbers based on the concepts of ETMMC are taken up, together with an integrated circuit (IC)-type absorber.

9.1 Basic Concepts and Means

As mentioned previously, ETMMC means that it equivalently converts the material constant using the same material or conventional materials, without having to produce new materials.

Electromagnetic Wave Absorbers: Detailed Theories and Applications, First Edition. Youji Kotsuka.
© 2019 John Wiley & Sons, Inc. Published 2019 by John Wiley & Sons, Inc.

The contents discussed in this chapter can be summarized as follows.

(a) The method of combining several materials in a checkerboard shape, using macro-size materials, rather than micro-size ones [4];
(b) The method of making small holes or slit perforations in the material [5, 6];
(c) The method of mounting periodic metallic patterns on the material surface;
(d) The method of attaching a conductive patch on the magnetic material surface [7];
(e) The method of attaching line conductive patterns on the magnetic material surface [8–12];
(f) The method of introducing a microwave integrated circuit concept [13, 14].

Especially, the last item (f) is extended to autonomous controllable metamaterials, as described in Chapter 10.

By introducing the idea of ETMMC, the following advantages may arise:

(1) Effective uses of the material can be achieved.
(2) When the characteristics are mismatched due to a measurement error in material constants or when the deviation of the dimensional accuracy are caused, the absorption characteristic can be easily improved.

The latter problem is particularly important when designing an EM-wave absorber in the high-frequency range, such as the terahertz band. For example, in 90 GHz, when designing the 1/4-wavelength EM-wave absorber, 1/4-wavelength becomes 0.75 mm. Even if a slight error occurs in this 1/4-wavelength dimension, the EM-wave-absorbing characteristic will have a much larger error with respect to the desired absorbing characteristics. In the following section, concrete examples of the absorbers based on the ETMMC idea are introduced.

9.2 Examples of ETMMC Absorbers

9.2.1 Microchip Integrated -type Absorber

The methods to realize the new EM-wave absorber characteristics by combining the existing EM-wave absorber materials are first considered [4]. The EM-wave absorber to be considered here is referred to as a "microchip integrated-type absorber," which is formed by integrating different kinds of minute chip-like materials. Because their chip sizes are on the order of millimeters, this EM-wave absorber is similar to an image of the parquet.

This configuration using rubber ferrite as absorber materials is shown in Figure 9.1. As shown in the figure, the absorber materials made of rubber ferrite A and B are arranged alternately like a checker board, and a group of

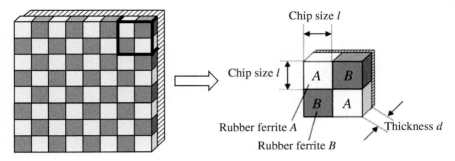

Figure 9.1 Structure of a ferrite microchip integrated-type absorber.

four chips is dealt as a unit cell. One side of the square chip is denoted by l, and the thickness of the ferrite is denoted by d.

Now, let us consider the absorber of the occupation area ratio A : B = 1 : 1 in each ferrite material for A and B, respectively, keeping the ferrite thickness a constant value of 6.2 mm. In this case, as is expected, the absorbing frequency of the present absorber takes an intermediate frequency between the frequencies that both ferrite materials of A and B possess. Under these conditions, the absorbing characteristics with simultaneous changes in both chip sizes l in the materials A and B ranging from 1 to 40 mm are shown in Figure 9.2. In this case, as is apparent from the figure, as the size of both square chips increases, the reflection coefficient level shows a tendency to degrade without changing its absorbing center frequency.

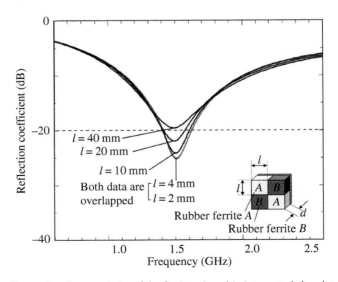

Figure 9.2 Absorption characteristics of the ferrite microchip integrated absorber.

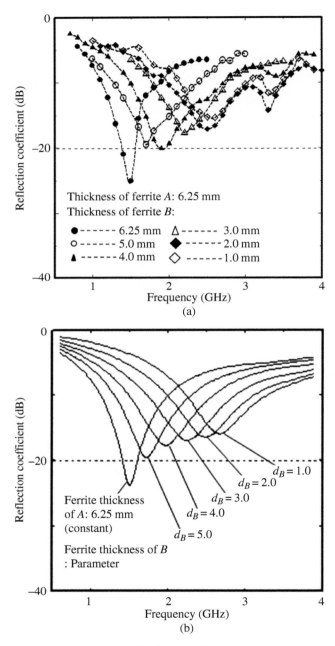

Figure 9.3 Absorbing characteristics of ferrite microchip integrated-type absorber. (a) Experimental results. (b) Analytical results.

Next, introducing the different ferrite thicknesses in materials A and B, let us investigate the change of absorbing characteristics in the case of designing a chip integrated-type absorber. While keeping each ferrite chip size constant and ferrite thickness of A at the constant value of 6.25 mm, the absorbing characteristics by both the measurement and theoretical analysis, when changing only the thickness of ferrite B, are shown in Figure 9.3. Each ferrite chip size is set to be a minute size, 1 mm in this case. From these measured and analytical values, it is found that the absorbing frequency shifts to the high-frequency region by reducing the thickness of the ferrite B.

Conversely, when reducing the thickness of ferrite A and keeping the thickness of ferrite B constant, almost the same characteristics are exhibited. As is evident from these examples, by combining the existing materials into a combination of various chips and changing the configuration conditions, it comes possible to adjust the central absorbing frequency and absorbing characteristics.

9.2.2 Absorber with Small Holes

In the rubber ferrite, it becomes possible to mainly change the dielectric constant in the material constants by punching out the variety of small holes, as shown in Figures 9.4 and 9.5 [5, 6].

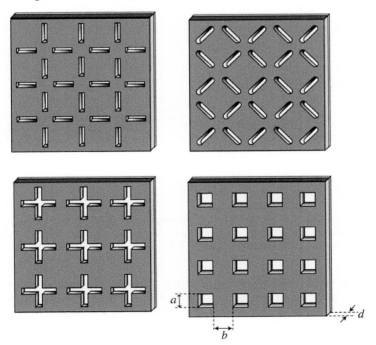

Figure 9.4 Examples of small hole shape for equivalent transformation of material constants.

Figure 9.5 Rubber ferrite with small holes. (Small holes are made by laser perforation.)

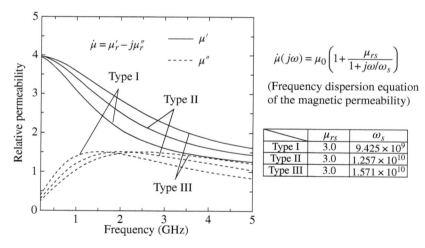

$$\dot{\mu} = \mu_r' - j\mu_r''$$

——— μ'

----- μ''

$$\dot{\mu}(j\omega) = \mu_0\left(1 + \frac{\mu_{rs}}{1 + j\omega/\omega_s}\right)$$

(Frequency dispersion equation of the magnetic permeability)

	μ_{rs}	ω_s
Type I	3.0	9.425×10^9
Type II	3.0	1.257×10^{10}
Type III	3.0	1.571×10^{10}

Figure 9.6 Complex relative permeability characteristics. (Type I in the figure shows measurement values. Type II and Type III represent theoretical values calculated from the frequency dispersion equation of permeability.)

In this section, let us investigate the EM-wave-absorbing characteristic in the case of providing the square holes in the rubber ferrite. As an example of the permeability characteristics, the frequency dispersion characteristics shown in Figure 9.6 are assumed. The relative dielectric constant of this ferrite material is 13.4. Type I exhibits frequency dispersion characteristics of the magnetic permeability, based on the measurement in the case of the existing Ni–Zn-based rubber ferrite. However, the permeability characteristics for Types II and III represent the theoretical values from the frequency dispersion equation depicted in the figure in order to grasp the following characteristics on the basis of the theoretical viewpoint. The solid lines in the figure represent the real part of the complex relative permeability, and the dotted lines show its imaginary part, respectively.

9.2.2.1 Effect of Square Hole Size

In this section, the topics are the absorber with square holes as an example of small holes. Also, the EM-wave-absorbing characteristics with respect to the constituent dimensions of this kind of small hole are discussed in detail from the theoretical analysis viewpoint. The result of the theoretical EM-wave-absorbing characteristics based on the finite difference time domain (FDTD) analysis, which was described in Chapter 4, is shown in Figure 9.7.

This figure represents the case where the size of the square holes is varied from 2.0 to 7.0 mm, keeping the thickness of the rubber ferrite and the spacing between adjacent holes constants at 6.5 and 4.0 mm, respectively. From the figure, it is confirmed that the EM-wave absorption frequency is shifted toward the high-frequency region, as the square hole size becomes large, when keeping the interval between holes *b* constant in Figure 9.4. This characteristic suggests that mainly the dielectric constant, rather than permeability, contributes to ETMMC.

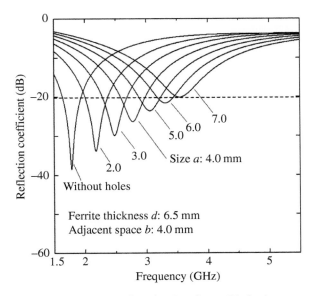

Figure 9.7 Absorption characteristics when the size of a small hole changes.

9.2.2.2 Effect of Adjacent Hole Space

Next, the theoretical absorbing characteristics are shown in Figure 9.8 in those cases where the interval b between the holes is narrowed from 8.0 to 1.0 mm, while keeping the thickness of the rubber ferrite and the size of the square holes constant at 6.5 and 2.0 mm, respectively.

Thus, by narrowing the adjacent interval b between the holes, the absorbing central frequencies can be shifted toward the high-frequency region. In the example of 8.0-mm spacing between holes, as shown in the figure, it is established that a good matching characteristic of approximately −40 dB is achieved at about 2.0 GHz. From these investigations, it is noticed that by providing the small holes to the ferrite absorber material, the permeability value, of course, also varies, but its change is less compared to the dielectric constant.

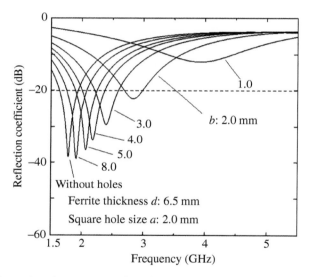

Figure 9.8 Absorption characteristics when the spacing between small holes changes.

9.2.2.3 Relation of Absorber Thickness and Hole Dimensions

Further, the mutual relationships between the rubber ferrite of thickness d and the hole size a in the case of the adjacent space constant 4 mm are investigated. Figure 9.9 exhibits the EM-wave-absorbing characteristics when the hole size a is varied from 7.0 to 1.0 mm by using ferrite thinned to 5.0 mm, compared to the ferrite thickness of 6.5 mm in the previous case. Similarly, Figure 9.10 shows the EM-wave absorption characteristics when the ferrite thickness d is 4.0 mm, taking the adjacent space b as a parameter.

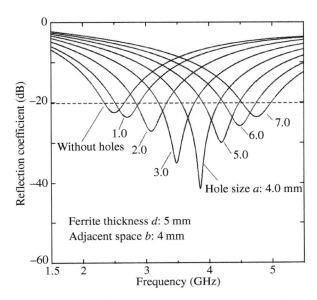

Figure 9.9 Relationship between the thickness of absorber material and the hole size. (Ferrite thickness is 5 mm.)

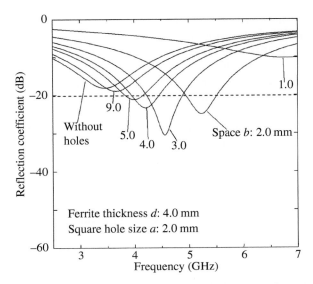

Figure 9.10 Relationship between the thickness of absorber material and the adjacent hole space. (Ferrite thickness is 4 mm.)

In this way, when thinning the ferrite thickness by increasing the hole size a and narrowing the adjacent space b, the tendencies of the EM-wave absorption frequency shifting to the higher frequency regions are the same as when the ferrite is thick.

This fact also implies that there is an optimal combination between the thickness of the ferrite material and the distance between adjacent holes. Although this ferrite material originally possesses the EM-wave absorption frequency at 1.8 GHz, it comes to maintain its absorption characteristics even in the high-frequency range such as several gigahertz by a simple means of providing holes using only a single ferrite material. These facts also suggest that the ferrite absorber can be made thinner in the higher microwave frequency regions than the matching thickness possessed by the original ferrite based on the simple method, described earlier.

9.2.3 Absorber with Square Conductive Elements

In this section, as another example of ETMMC, a case where a square thin metal plate is periodically arranged on the surface of the EM-wave absorber to change the absorber characteristic is described in detail [7]. By mounting the conductive elements of various shapes on the absorber surface, the absorbing frequency characteristics, present originally, can be shifted into both the low- and high-frequency ranges. By introducing this means, it also becomes possible to improve the thinning of the absorber and its absorbing characteristics itself. It should be noted that the dimensions of the conductor elements adopted here are sufficiently smaller than those of the wavelength.

Hereafter, mainly the characteristics of the absorber, having the square periodic conductor element on its surface, are described. Figure 9.11a shows a configuration example when a square conductive element is mounted on the surface of an absorber. Figure 9.11b shows a photograph when a square aluminum element is actually attached to the rubber ferrite surface. The intention of introducing this form is to add capacitance and inductance to the EM-wave absorber material by mounting various types of conductive elements on its surface. As a result, both the absorber frequency characteristic and the absorbing characteristic of the absorber can be changed relatively easily by adjusting the dimensions of these conductive elements and their mutual spacing [7, 8]. For reference, the data obtained by comparing the experimental values with the analytical values, when square conductors are periodically arranged on the surface of the rubber ferrite EM-wave absorber, is shown in Figure 9.12b.

Figure 9.13 shows the frequency dispersion characteristics of the permeability of rubber ferrite in this case. The relative permittivity of this rubber ferrite is $5.6-0.2j$. The size of one side of the square conductive element is 10.0 mm, which is made of thin aluminum, and both mutual distances between the conductive elements are 10.0 mm. The FDTD method was also introduced in the present analysis.

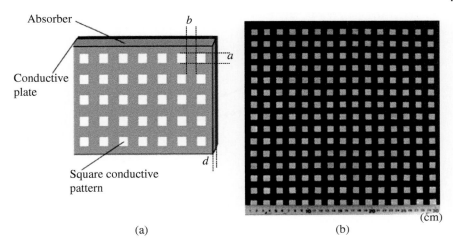

Figure 9.11 Examples of EM-wave absorber with conductive square frame patterns. (Gray color: magnetic material, white: conductive square element, *a*: square element size, *b*: adjacent space, *d*: ferrite thickness). (a) Configuration. (b) Picture.

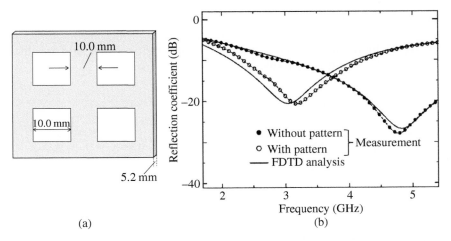

Figure 9.12 Comparison of analytical and measured results of absorbing characteristics in an absorber with square frame patterns. (a) Absorber with square frame conductive patterns. (b) Absorber characteristics with square conductive frame patterns. (The size *a* of one side of the periodic conductor square element made of thin aluminum is 10.0 mm, and the spacing *b* between elements is 10.0 mm in this case. The thickness of the dielectric sheet supporting the conductive element is 0.15 mm. Also, the value of the dielectric constant of the rubber ferrite material is 5.6–0.2*j*.)

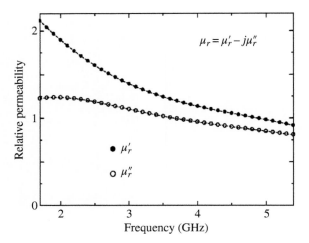

Figure 9.13 Permeability characteristics of rubber ferrite materials.

9.2.3.1 Effect of Conductor Dimensions

Here, the effect on the absorbing characteristic between the size of the conductor element and the adjacent distance b in the present absorber configuration, as shown in Figure 9.12a, are investigated. First, the absorbing characteristics of rubber ferrite when the conductor dimension a is kept constant and the adjacent space b is taken as a parameter, is shown in Figure 9.14.

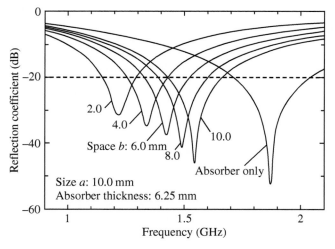

Figure 9.14 Analytical results of absorption characteristics when the distance b between conductors changes.

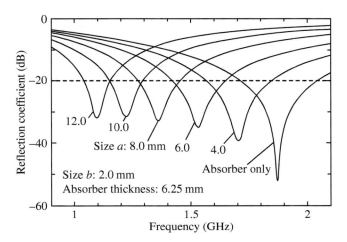

Figure 9.15 Analytical results for absorption characteristics when the size of the conductor element changes. (The permeability value of the rubber ferrite material is used in this case. Its relative permittivity is 14.0. The conductivity of aluminum as the periodic conductor elements is 3.546 × 107 S/m. The thickness of the dielectric film on which the periodic conductive element is deposited is 0.1 mm, and its relative permittivity is assumed to be 1.0.)

In Figure 9.14, when the conductor dimension a has a constant value of 10.0 mm and the conductor spacing b is reduced from 10.0 to 2.0 mm gradually, it is found that the absorber characteristic gradually deteriorates and moves toward the low-frequency side.

Next, Figure 9.15 represents the absorbing characteristics when the adjacent space b between the conductive element is constant, and the conductor element size a is changed. From Figure 9.15, it can be found that the absorption frequency moves toward the lower frequency side, while the adjacent distance b between conductive elements is kept constant at 2.0 mm, and as the conductor element size a gradually increases from 4.0 to 12.0 mm. In short, the absorber center frequency shifts to the low-frequency region by increasing the conductor element size a or decreasing the adjacent space b between the conductive elements. From these facts, it is found that the size of the conductor element a and the adjacent distance b are the main constituent parameters that determine the absorption characteristics of the present EM-wave absorber.

By the way, as is described later, one of the characteristics of this mounting method is that it becomes possible to reduce the absorber thickness using a magnetic material with a large dielectric constant [11]. Carbonyl iron material is one of the candidate materials satisfying this condition. The results of the absorption characteristics analysis when using carbonyl iron materials are shown in Figure 9.16a,b. This figure represents a case where the size of the

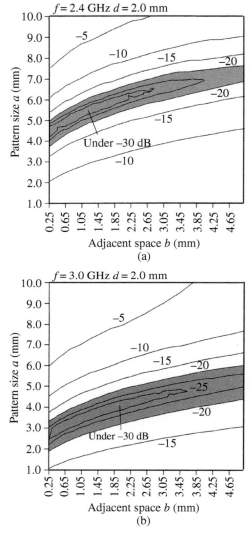

Figure 9.16 Relationship between the size of the square conductor frames and the spacing between conductor frames (absorption characteristics when the size *a* of the periodic conductive frame and the adjacent distance *b* between them varies in different ways. (a) Focuses on the absorption frequency of 2.45 GHz, and (b) focuses on the case of the absorption frequency of 3.0 GHz). These results also suggest that the best combination exists between the size of the conductor element *a* and the adjacent distance *b* between the conductor elements.

periodic conductor element *a* and adjacent space *b* are changed variously as the parameters.

Numerical values in the figure indicate the value of the reflection coefficient (dB). These figures mean that the area of light gray color is below −20 dB in reflection coefficients.

Also, from Figure 9.16a, it can be noted that even at a frequency of 2.45 GHz the absorber thickness *d* is only 2.0 mm, sufficient EM-wave absorption characteristic yields in this thickness.

9.2.3.2 Input Admittance Characteristics

Now, let us consider how the wave absorption characteristic can be changed by attaching periodic conductive elements to the absorber surface from the viewpoint of the behavior of the input admittance looking toward the absorber direction. Again, note that we always assume a conductor size smaller than the wavelength. First, Figure 9.17 shows the analysis result of the input admittance when the conductor element width *a* is kept constant and the adjacent distance *b* is changed.

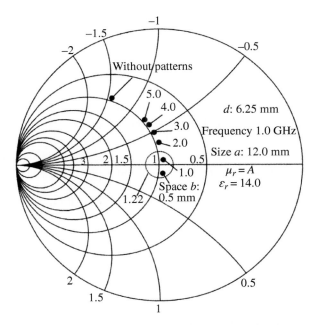

Figure 9.17 The input admittance behaviors on the Smith chart when looking into a termination direction from absorber surface. (Adjacent distance *b* is taken as a parameter.)

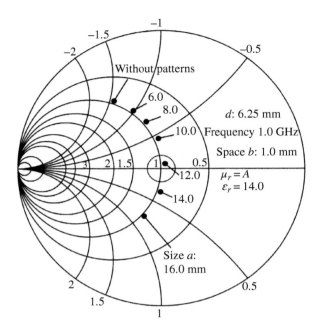

Figure 9.18 Input admittance characteristics. (Conductor width *a* is taken as a parameter.)

Figure 9.17 shows that the susceptance component of the input admittance increases by narrowing the space *b* between adjacent conductor elements, and the susceptance value shows the best matching characteristic in the vicinity of *b* = 1.0 mm. Similarly, Figure 9.18 shows the input admittance characteristics when the adjacent space *b* is assumed to be a constant value and the conductor width *a* is taken as a parameter. By increasing the size of the conductor *a*, the susceptance component increases, and good matching characteristics are attained when the size of the conductor is in the vicinity of 12.0 mm in the present case. From these data, it can be found that the input admittance of the ferrite absorber takes inductive values when it is sufficiently thin compared to the wavelength. Therefore, in order to obtain the matching characteristic of this kind of magnetic absorber, it may be configured so as to cancel the susceptance component of the input admittance, which the magnetic absorber inherently possesses. This is made possible by attaching a conductive element to the surface of the EM-wave absorber to impart capacitive properties.

Considering this input admittance, it is also noticed that the periodic conductive element provided on the front surface of the EM-wave absorber operates as the impedance conversion layer. In this regard, refer to the investigations in Ref. 11, which applied the equivalent circuit model based on the transmission line theory.

9.2.4 Absorber with Line-Shaped Conductive Elements

In this section, let us introduce the effectiveness of ETMMCs in the case of introducing periodic conductive line frames (PCLFs) [8–10]. The aim of introducing the PCLFs is that the absorber characteristics are expanded more, compared with the previous case of square conductor element. This is because the effect of L, C, and R components, representing the equivalent circuit of the absorber, makes the absorber characteristics control easier as compared to the case of the square conductor elements. Various shapes of PCLFs associated with this section are shown in Figure 9.19. Figure 9.19a,b illustrates patterns with a thin line lattice and a cross, respectively. Figure 9.19c shows a structure in which the tip of the pattern of the cross shape is inserted in the surface portion of the ferrite. Figure 9.19d represents the case of a square line pattern. Figure 9.19e denotes double-layered PCLFs. The back surface of these EM-wave absorbers, made of magnetic materials, is attached to the conductive plate.

As for the magnetic EM-wave materials in each of these configuration methods, we have to select a material which can attain optimal characteristics among the candidates made of sintered ferrite, rubber ferrite, carbonyl iron, or the like.

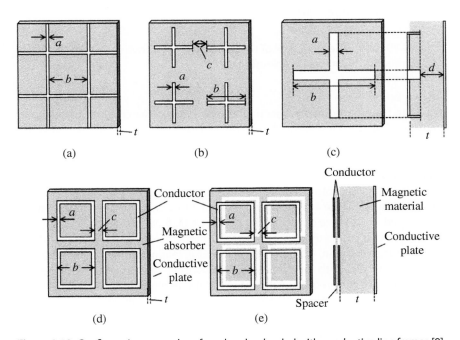

Figure 9.19 Configuration examples of an absorber loaded with conductive line frames [9]. (a is the width of the conductive line pattern mounted on the surface layer, b is the distance between the conductive line patterns, and t is the thickness of the absorber.) (a) Lattice type, (b) cross type, (c) insert type, (d) square frame type, and (e) double-layered type.

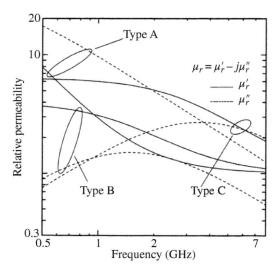

Figure 9.20 Values of relative permeability (A: sintered ferrite, B: rubber ferrite, C: carbonyl iron). (Relative permittivity values of A: 12.0, B: 14.0, C: 20.0, respectively.)

Figure 9.20 shows the examples of the permeability characteristics of the material adopted in the present implementation.

9.2.4.1 Lattice Type

As pointed out in the previous section, the idea of a square conductive-element-type absorber is designed to add electrostatic capacitances using square conductor elements on the surface of the absorber that is inductive in nature (Figure 9.11). In other words, this method was based on the principle of absorbing the incident wave by canceling the susceptance components with each other. On the other hand, in the present case of the lattice type, the contribution of charge accumulation is small and a current is generated on the line-shaped conductive element. The magnetic field, which is generated by this conductive current, causes an opposite magnetic field component against the incident magnetic field, that is, a demagnetizing magnetic field. As a result, the equivalent permeability of the EM-wave absorber decreases, and, therefore, the matching center frequency moves to the high-frequency region. This is the operating principle of the lattice-type conductive frame absorber. With this in mind as the difference from the other patterns, let us investigate the absorber characteristics of the present case of the lattice-shaped conductor element shown in Figure 9.21b.

Let us show the absorbing characteristics in the case of taking the distance b between the mutual lattice conductive lines as a parameter in Figure 9.21. From this figure, by keeping a thin constant line lattice width $a = 1.0\,\text{mm}$, and gradually narrowing the adjacent conductor line interval b, it is found that the original

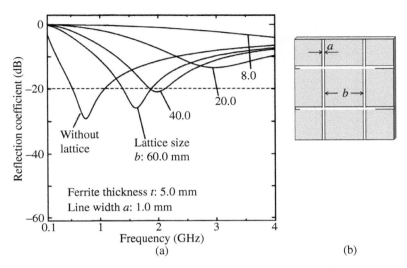

Figure 9.21 (a) Wave absorption characteristics in a lattice type. (As the material constant in the present absorber, the dielectric constant value is 12.0, and the value of sintered ferrite is adopted as the relative magnetic permeability (Type A).) (b) Dimensions of lattice type absorber.

center matching frequency of the absorber moves gradually to the higher frequency region.

9.2.4.2 Cross Type

Figure 9.22a shows the comparisons of theoretical analyses with actual measurements when a cross-shaped conductor shown in Figure 9.22b is attached to

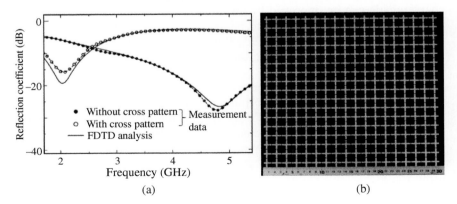

Figure 9.22 (a) Absorber characteristic with cross patterns. (b) Actual appearance of prototype absorber. (As the relative magnetic permeability, a value of the rubber ferrite material shown in Figure 9.22 (Type B) is used, and the relative dielectric constant is 5.6–0.2j and thickness is 5.2 mm. The line width a of the conductive cross frame is 2.5 mm, the size b of the element is 16.0 mm, and the spacing c between the elements is 1.0 mm.)

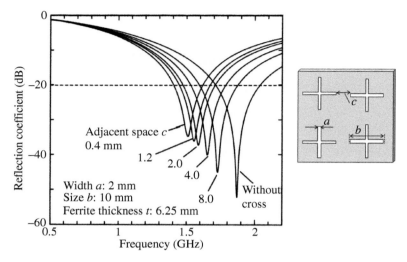

Figure 9.23 Absorbing characteristics when the tip of the cross-shaped element is inserted into the EM-wave-absorbing material.

the surface of the absorber. As material constants of the EM-wave-absorbing material, the relative permittivity is 14.0, and the relative permeability is assumed to be the same as in the rubber ferrite material in Figure 9.20 (Type B). As shown in Figure 9.23, by narrowing the adjacent distance c between the cross-shaped conductive elements, the EM-wave absorption center frequency can be changed to the low-frequency side, but its changing amount is small. Regarding this characteristic, since it is largely dependent on the element sizes between adjacent cross-shaped conductive elements, this characteristic can be improved by introducing the architecture of the tip of the cross-shaped element so as to be inserted into the EM-wave-absorbing material, as shown in Figure 9.19c. This is because the capacitive effect is increased compared to the configuration case in Figure 9.19b.

9.2.4.3 Square Conductive Line Frame
In this section, the EM-wave absorption mechanism in the case of a periodic square conductive line frame in Figure 9.19d is described in detail. It should be noted here that the present theoretical considerations are described on the basis of predetermined assumptions. This prerequisite condition is that we have to preselect a magnetic absorber material that can match without using the PCSFs at a certain frequency and with some absorber thickness. That is, from the Smith chart viewpoint, this means that the normalized input admittance (NIA), which is looking toward the magnetic absorber surface, has a value of 1.0 in its normalized input conductance value at some frequency and at some absorber thickness. In the given investigations, as the

present material constants, the permeability values, which have the frequency dispersion characteristics shown in Figure 9.20 (Type B), are adopted. The relative permittivity is 14.0.

Effect of Square Frame Line Width Let us first investigate the absorber characteristics when only a conductive square frame width a is varied, keeping the other configuration dimensions as constants. As mentioned, notice again that the prerequisite condition is imposed beforehand in that the NIA looking toward the EM-wave absorber takes a value of 1.0 on the Smith chart at some frequency and for some absorber thickness. This means that we have to choose the magnetic absorber material which can satisfy the absorbing condition of the EM-wave absorber at some frequency and in some absorber thickness without mounting the PCLFs. The analytical result of the EM-wave absorption characteristics in this case is shown in Figure 9.24.

In the present case, this absorber satisfies the prerequisite condition with the normalized input conductance value of 1.0 at the frequency of 1.85 GHz when the absorber thickness is 6.25 mm. As for the square frame conductivity, a value of 3.5×10^7 S/m for aluminum is used. In addition, the size of the square frame b is 12.0 mm, and the adjacent space c is 2.0 mm. It has again been found that the absorbing frequency characteristic is shifted toward a lower frequency than the original absorbing frequency at 1.85 GHz as the width a is increased. The conductive line frame eventually coincides with the square conductive patch, when the width of the conductive line frame is 6.0 mm.

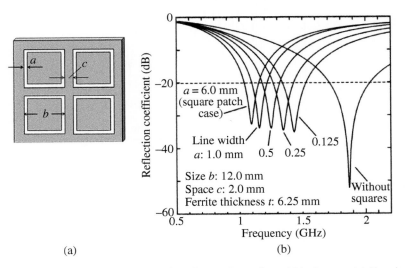

(a) (b)

Figure 9.24 Absorber characteristics when the conductor line width changes. (a) Absorber with the square frame lines. (b) Absorber characteristics when taking the conductor line width a as a parameter.

Effect of Square Frame Width Next, Figure 9.25 shows the absorber characteristics when the square frame size *b* is taken as a parameter, the other parameters being constant. The width *a* and the space *c* are 1.0 and 2.0 mm, respectively. The size *b* is varied from 4 to 22 mm. From Figure 9.25, it is found that the absorbing frequency characteristic tends to move toward a lower frequency as the frame size *b* is increased. This absorbing phenomenon can be explained by paying attention to the capacitance change between the adjacent spaces of the PCLF, as described before. In addition, if the EM-wave absorber part is represented by an equivalent transmission line circuit, the present function of PCLFs can be explained as a phenomenon in the resonance circuit caused by a capacitor connected in parallel to this transmission line. Concerning the absorbing characteristics in Figure 9.25, it can also be pointed out that the EM-wave-absorbing characteristic shows a tendency with a sharpness in the low-frequency region from the behavior of the quality factor value *Q* in the equivalent transmission line. For a detailed discussion of this issue, refer to the literature [11].

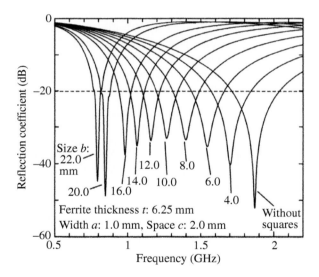

Figure 9.25 Absorption characteristic when the size *b* of a square line frame changes.

Effect of Adjacent Space

Next, let us investigate the effect of adjacent space *c* of PCLFs.

Figure 9.26 represents the theoretical analysis results of the present absorber characteristics when other parameters are kept constant, assuming the adjacent space *c* between PCLFs as a parameter. In the present case, the size *b* and the line width *a* are 12.0 and 1.0 mm, respectively.

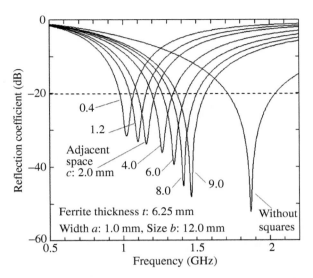

Figure 9.26 Absorber characteristics with adjacent space *c* as a parameter (ε_r : 14.0, μ_r : Type B).

It can be confirmed that the EM-wave-absorbing frequency characteristic is shifted toward a lower frequency region than the original matching frequency at 1.85 GHz, as shown in Figure 9.26, if the adjacent space *c* is decreased.

Next, in order to investigate the present absorber frequency characteristic from the viewpoint of NIA on the Smith chart, let us consider the present absorber characteristics based on the FDTD analysis. When the adjacent interval *c* of the PCLF is taken as a parameter, the characteristics of the NIA on the Smith chart, which is looking toward the surface of the EM-wave absorber loaded with the PCLFs, change as shown in Figure 9.27. The spacing *c* takes the values from 10.0 to 0.4 mm at each frequency of 0.8, 1.0, 1.5, 2.0, and 2.5 GHz.

The symbols (⊙◎♠◌●) represent the NIA values in the case without mounting PCLFs on a rubber ferrite absorber surface, and these symbols correspond to a frequency of 0.8, 1.0, 1.5, 2.0, and 2.5 GHz, respectively. From this chart, it can be found that the normalized admittance values (●,○,△) at each frequency from 0.8 to 1.5 GHz rotate clockwise around the circle with an NI conductance value of 1.0 by decreasing the adjacent space value *c* from 10 to 0.4 mm [10].

As is seen from the study of the Smith chart, the effect of decreasing the adjacent space *c* is to make the present susceptance value approach toward the conductance value of 1.0. Thus, we can find that the present absorbing condition is based on this kind of mechanism. Note that this method can be applied in lower frequency regions than the original matching frequency of 1.85 GHz of the present rubber ferrite absorber in the case without PCSFs.

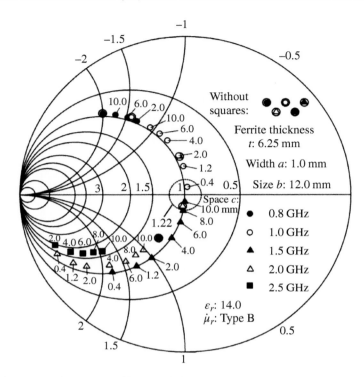

Figure 9.27 Normalized input admittance (NIA) with space *c* as a parameter.

9.2.4.4 Double-Layered PCLF Type

Next, the absorber frequency characteristics in double-layered type of SCLFs (Square Conductive Line Frames) are investigated. The present ferrite under investigations is a rubber ferrite having a thickness of 6.25 mm and having magnetic permeability characteristics in Type B from Figure 9.22. The medium between the two PCLF substrates is made of a dielectric material with a relative permittivity of 2.0, and its thickness is 1 mm. In addition, as shown in Figure 9.19(e), the configuration dimensions in this case are 4.0, 12.0, and 16.0 mm in the line width *a*, the pattern width *b*, and the adjacent space *c*, respectively. Figure 9.28 shows the EM-wave absorption characteristics of the double-layered PCLF-type absorber in the present case. As can be seen from this figure, in the case of a two-layered PCLF absorber, two absorbing frequencies can be realized.

Next, from the viewpoint of further deepening our knowledge of the EM-wave absorber characteristics, when this kind of conductive element is loaded, we try to investigate from the viewpoint of the NIA by introducing the Smith chart. Figure 9.29 shows the NIA characteristics in the case of a

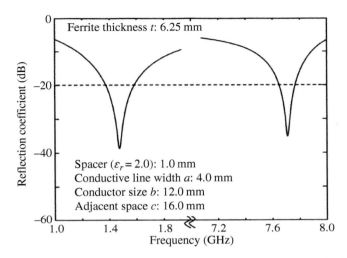

Figure 9.28 Theoretical analysis of matching characteristics of a double-layered absorber in Figure 9.19e.

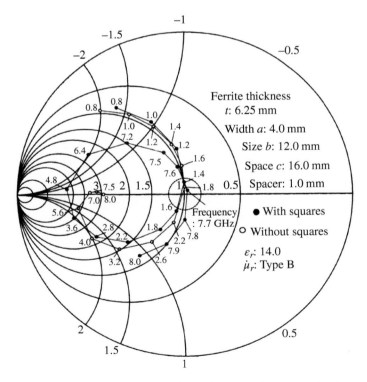

Figure 9.29 NIA characteristics in the case of a double-layered-type absorber in Figure 9.19e.

double-layered PCLF-type absorber. In this Smith chart, the NIA values in cases with and without PCLFs are represented by taking the frequency as a parameter. Since Figure 9.29 is somewhat complicated, let us further explain its meaning in this Smith chart. By introducing the equivalent transmission line to this absorber of the PCLF configuration, it becomes clear that both the susceptance and the conductance on the Smith chart can be controlled.

As shown in Figure 9.29, it is noticed that the NIA value (●) keeps a circle shape and again returns to the EM-wave-absorbing area when the frequency shifts to a higher value. Referring to the present Smith chart representation, we can also investigate the twin-peak matching mechanism based on the equivalent transmission line theory, related to this kind of absorber. Concerning the detailed explanations for this, refer to Ref. [10]. In this way, the NIA behaviors and the absorbing characteristics can be understood together with their physical images from the standpoint of the equivalent transmission line theory by introducing the Smith chart. These characteristics and functions of the EM-wave absorber, which introduce the periodic conductive elements, are summarized in Table 9.1.

Table 9.1 Characteristics and functions of the EM-wave absorbers with periodic conductive elements.

Conductive pattern shapes	Special features	Functions
	Capacitance value becomes large. Even if patch size is small, matching characteristic can be changed largely	Change to lower frequency
	By adjusting inductive susceptance component, matching characteristics are changed to a higher frequency region	Change to higher frequency
	Capacitance value is small compared to that of another pattern	Change to lower frequency
	Matching central characteristic can be changed finely, particularly in the millimeter frequency region	
	If the frame size satisfies the condition of $a/b > 0.0833$, this pattern shows the same characteristic as that of the square patch case	Change to lower frequency
	Dual matching characteristics can be realized when adjusting susceptance at low frequency and both susceptance and conductance at a higher frequency region	Dual matching characteristics

Finally, let us suggest how to make the absorber materials. When making the holes, the hole sizes are relatively small, so accurate size is required. In this regard, it is important to adopt rigorous fabrication method. In order to periodically make a conductive element composed of a thin thickness on the surface of the electromagnetic (EM)-wave-absorbing material, the means for vacuum deposition on should be introduced. Further, although only the magnetic material is considered here, the ETMMC method can also be applied to other materials, including lossy dielectrics. From these viewpoints, it may be concluded that with the recent various kinds of technological progress, it has become possible to manufacture the absorbers by the ETMMC methods for the first time.

9.2.5 Absorber Based on Integrated Circuit Concept

Conventionally, the magnetic material represented by ferrite, the carbon-based material, and the dielectric material, and the like have been mainly used as the absorber materials. When these absorber materials are used, as already explained, the absorber characteristics are limited by the characteristics of the permeability, permittivity of the material, and construction shape, and each dimension of absorber that will be realized. In the future, the design of the absorber will be demanded in a high-frequency band such as the millimeter and terahertz waves. Regarding this point, as is clear from the Snoek's limit of the ferrite, the use of magnetic materials becomes more difficult in the high-frequency region.

To break through this kind of problem, let us introduce a new EM-wave absorber which does not depend on the frequency dispersion characteristics of materials. This absorber is composed of the integrated circuit concept [13, 14].

The EM-wave absorption principle of this absorber can be easily explained by replacing it with a microwave frequency circuit. That is, after introducing the impedance matching circuit that cancels the reactance component of the input impedance, an absorber can be designed to efficiently dissipate the high-frequency energy using only the passive resistor component. Accordingly, it is only by changing the circuit configuration elements, such as the passive circuit components and the substrate structure forms, that the various EM-wave-absorbing characteristics can be freely realized. Since this absorber can be considered, in a broad sense, as an EM-wave absorber that applies the ETMMC concept, this absorber is considered here. For this method of implementation, this EM-wave absorber is characterized by a nature of thin thickness and light weight, even with the characteristic of a broadband.

9.2.5.1 Configuration of Absorber

A basic configuration example of this EM-wave absorber is shown in Figure 9.30. This configuration is composed of a substrate-mounted conductive circuit element with microchip resistors and a spacer between the substrate

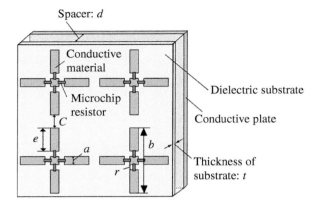

Figure 9.30 Integrated circuit–type absorber compatible with vertical and horizontal polarized waves.

and a plate of conductive terminal end. The characteristics of the absorber are governed by the resistance value of the microchip resistor, the size of the configuration of the conductive circuit element in each part, and the spacer distance d. Among them, the means for strictly controlling the resistance value of the microchip resistor is contrary to the present absorber design policy. Therefore, the method of realizing easily the absorber characteristic which we want must be investigated. In the following, let us investigate how to solve this kind problem, using FDTD analysis.

Now, the absorbing center frequency f_0 and the relative bandwidth BW of the EM-wave absorber are defined by the following expressions.

$$f_0 = \frac{(f_H + f_L)}{2} \ (\text{Hz})$$

$$BW = \frac{(f_H - f_L)}{f_0} \times 100 \ (\%).$$

Here, f_H is the upper limit frequency satisfying the require reflection coefficient and f_L is the lower frequency.

In Figure 9.31, all parameters of the structural dimensions are normalized by the wavelength of the central absorbing frequency presenting a good absorbing characteristic. In the present case, the maximum band width of 55% can be obtained if the required reflection coefficient is -20 dB in the case of ④. This central absorbing frequency is 3.5 GHz. Therefore, all structural dimensions are normalized by this wavelength λ_0 at 3.5 GH in the following investigations. Notice here that the normalized resistance values of the microchip resistors are represented by the normalized resistivity value ρ using the actual wavelength λ_0 without using the resistance value.

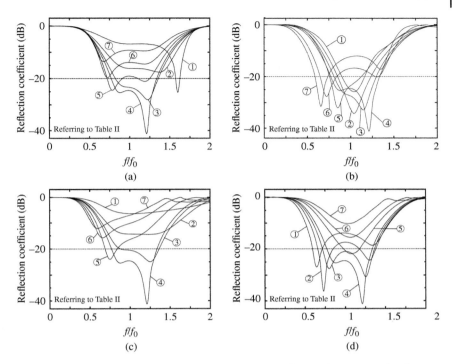

Figure 9.31 Analysis results of absorption characteristics when each parameter is taken as a parameter. (a) The case of taking normalized resistivity ρ by the wavelength of the central absorbing frequency as a parameter. In the present case, the maximum bandwidth is 55% (figure of merit) if the required reflection coefficient is −20 dB. This is the case of ④ in (a). And the central absorbing frequency is 3.5 GHz. These values are applied throughout the following discussions. (b) The case of changing the line width a and a conductive line pattern length b. (c) The case of changing the length e and the length b simultaneously. (d) The case of changing the adjacent distance c between the conductive circuit elements. In all these investigations, the normalized resistivity ρ is kept in the case of ④ in (a).

Further, in these figures, the frequency axis as the horizontal axis is normalized by the center frequency f_0 in the band where the reflection coefficient takes the value of −20 dB. All these structural parameters (a, b, c, d, e, r) including the normalized resistivity ρ are tabulated in Table 9.2.

From these investigations, it becomes possible to suggest that the desired mono-peak or twin-peak absorbing characteristics can be realized over a wideband by adjusting the size of the high-frequency circuit dimensions, even if the microchip resistance value cannot be strictly defined. Figure 9.32 represents the absorbing characteristics when taking the actual resistance value as a parameter. According to Table 9.2, the resistance values are 75 or 82 Ω in case of ④ and ③, respectively, and it is recognized that the broadband characteristic in

Table 9.2 Example of each parameter normalized by the wavelength of the center absorbing frequency.

	Figure 9.31a; ρ: parameter	Figure 9.31b; a,b: parameter		Figure 9.31c; e,b: parameter		Figure 9.31d; c: parameter
	ρ	a	b	e	b	c
①	0.095	0.012	0.385	0.070	0.210	0.012
②	0.048	0.023	0.397	0.140	0.350	0.023
③	0.039	0.035	0.409	0.163	0.397	0.035
④	0.035	0.047	0.420	0.175	0.420	0.047
⑤	0.030	0.058	0.432	0.210	0.490	0.093
⑥	0.024	0.105	0.479	0.245	0.560	0.140
⑦	0.018	0.140	0.514	0.280	0.630	0.234
a	0.047	Parameter		0.047		0.047
b	0.420	Parameter		Parameter		0.420
c	0.047	0.047		0.047		Parameter
e	0.175	0.175		Parameter		0.175
r	0.012	0.012		0.012		0.012
d	0.198	0.198		0.198		0.198
ρ	Parameter	0.035		0.035		0.035

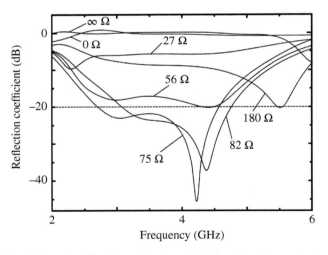

Figure 9.32 Analysis result of the absorption characteristics when the spacer size is changed.

the reflection coefficients with less than −20 dB can be realized [14]. This also suggests that it is not necessary to strictly control the resistance value in this EM-wave absorber configuration.

By the way, the purpose of constructing this EM-wave absorber is to establish a design method that ensures a consistent configuration up to the high-frequency range. From these theoretical investigation results, it can be deduced that desired mono-peak or twin-peak absorbing characteristics can be realized over a wideband by only adjusting the size of the high-frequency circuit even if the microchip resistance value cannot be strictly defined. The idea in the background of applying this IC-type EM-wave absorber to the high-frequency band design comes from these facts.

9.2.5.2 Space Experiment Characteristic

Let us introduce the experimental result in an anechoic room here. Figure 9.33 shows the photograph of the actual IC-type absorber for use in the measurement.

Figure 9.34 represents a comparison of the results under analysis condition of ④ with the resistance value of 75 Ω shown in Table 9.2, and the measured values in the case of normal incidence. The spacer thickness is 17 mm. These absorbing characteristics show a normal incident wave case.

But the oblique incidence characteristics are about −15 dB for an angle of incidence of 35° and about −10 dB for 45° in both the TE and TM waves.

This IC-type absorber has the property that it can be used even in a higher frequency region, such as the terahertz region, that does not depend on the absorber material constants and have wideband characteristics and light weight. In such a higher frequency region, the circuit size of the absorber

Figure 9.33 A photograph of an IC type prototype absorber.

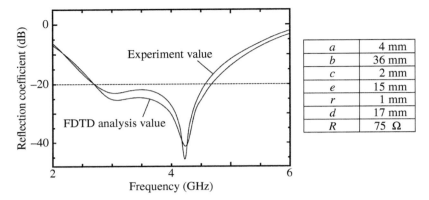

a	4 mm
b	36 mm
c	2 mm
e	15 mm
r	1 mm
d	17 mm
R	75 Ω

Figure 9.34 Comparison between the theoretical analysis and measured values.

becomes very small. When designing in this high-frequency region, the integrated circuit technology is required. That is why this absorber is called an integrated circuit-type absorber. Here, although the cross shape is selected as a conductor circuit element of this IC-type EM-wave absorber, it is noted that a circuit element of an arbitrary shape can be applied as a matter of course.

References

1 Kotsuka, Y. and The Latest EM-wave Absorber Technologies (2004). EM-wave absorber based on equivalent transformation method of material constants. In: *IEICE Society Conference, BS-1-1*, S-1–S-2.

2 Kotsuka, Y. (2005). Editorial preface & overview in special issue: "On advanced technologies of EM-wave absorber enhancing ubiquitous network environment". *J. IEICE* 88 (12): 931–935.

3 Kotsukas, Y. and Shimodaira, K. (2006). A theoretical approach to matching characteristics of a novel absorber based on the concept of equivalent transformation method of material constant. *IEICE Trans. Electron.* E89-C: 2–8.

4 Nagoya, S., Amano, M., and Kotsuka, Y. (2001). On the matching characteristics of the wave absorber composed of the ferrite minute chip. In: *Technical Report of IEICE, EMCJ-9.3*, 91–96.

5 Kotsuka, Y. and Amano, M. (2001). A method of effective use of ferrite for microwave absorber. In: *IEEE MTT-S International Microwave Symposium Digest*, 1187–1190.

6 Amano, M. and Kotsuka, Y. (2003). A method of effective use of ferrite for microwave absorber. *IEEE Trans. Microwave Theory Tech.* 51 (1): 238–245.

7 Amano, M., Katsuta, K., and Kotsuk, Y. (2001). Matching characteristics of thin EM-wave absorber with periodically distributed conductors. In: *Technical Report of IEICE, EMCJ2001-52*.

8 Kotsuka, Y., Amano, M., Katsuta, K. et al. (2001). The FDTD analysis of the EM-wave absorber with conductor on the ferrite surface. In: *National Convention of IEICE (Japan)*, No. B-4-66, 381.

9 Amano, M. and Kotsuka, Y. (2002). A novel microwave absorber with surface-printed conductive line patterns. In: *2002 IEEE MTT-S International Microwave Symposium Digest*, 1193–1196.

10 Amano, M. and Kotsuka, Y. (2015). Detailed investigations on flat single layer selective magnetic absorber based on the equivalent transformation method of material constants. *IEEE Trans. Electromagn. Compat.* 57 (6): 1398–1407.

11 Katsuta, K., Amano, M., and Kotsuka, Y. (2001). Matching characteristics of ferrite absorber using stratified conductive short-plate. In: *Technical Report of IEICE, EMCJ2001-53*, 7–12.

12 Amano, M. and Kotsuka, Y. (2003). Matching characteristics of EM-wave absorber with metallic patterns on the front and back surface. *IEICE Trans.* J86-B (7): 1165–1175.

13 Kotsuka, Y. and Amano, M. (2003). Broadband EM-wave absorber based on integrated circuit concept. In: *2003 IEEE MTT-S International Microwave Symposium Digest*, vol. 2, 1263–1266.

14 Kotsuka, Y. and Kawamura, C. (2005). Proposal of a new EM-wave absorber based on integrated circuit concept. *IEICE Trans.* J88-C (12): 1142–1148.

10

Autonomous Controllable-Type Absorber

The rapid progress of artificial intelligence (AI) technology is remarkable.

With its rapid progress, it seems that development of whole technological fields needs to be considered into while facilitating their association with AI technologies in their thinking about development hereafter.

Chapter 9 described how to construct electromagnetic (EM)-wave absorbers on the basis of the equivalent transformation method of material constants (ETMMC) idea as a new absorber configuration concept. This chapter introduces a new concept of "autonomous controllable metamaterial" (ACMM) that can be assimilated with AI technology and various characteristics when applying a part of this material as an EM-wave absorber. This absorber can correspond not only to both TE-wave and TM-wave polarization problems but also to electrical controllable EM-wave absorbers that can satisfy all the conditions needed in an EM-wave absorber. Section 10.1 describes the necessity of this new metamaterial, its proposal background, and its configuration method. In Section 10.2, the implementation methods of ACMM-type absorbers are described. In Section 10.3, the conditions that an absorber should have are summarized, and the methods of realizing these absorbing conditions are investigated from every angle. It is important to know the EM-wave absorber behaviors in an ACMM-type absorber. For this purpose, input impedance investigations are conducted in Section 10.4.

10.1 Autonomous Control-type Metamaterial

As is well known, it was said that all science and technology are oriented toward "ecosystems" [1]. From this source, the concept of an ACMM arose. ACMM, which is introduced in this chapter, is defined as those that fused the nature of a living cell and the structure concept in a crystal lattice. In other words, the implementation method of ACMM has been proposed on the basis of the following two different concepts [2–7].

Electromagnetic Wave Absorbers: Detailed Theories and Applications, First Edition. Youji Kotsuka.
© 2019 John Wiley & Sons, Inc. Published 2019 by John Wiley & Sons, Inc.

The first is the biological concept. Generally, a living organism has a function called "homeostasis" that "autonomously" maintains a steady state at all times against blood pressure, sudden fluctuations in blood glucose levels, and the like. Therefore, the present material is named after this biological function. This concept can contribute to the protection from certain adverse electromagnetic circumstances by means of detecting them beforehand, using the sensor and material function which can absorb the EM wave autonomously, like in a biological function. The implementation concept of this material can be assimilated with the AI technology itself with the help of such a function.

The second concept is associated with material configurations. This is based on referring to the structure of the Bravais crystal lattice, which defines the formation of a solid material. Referring this norm will become important to suggest an optimum unit cell arrangement to this material when the arrangement becomes more complex in the future. As shown in Figure 10.1, the ACMM structure consists of the following components to realize the abovementioned concepts.

(a) An active circuit element which electrically controls electric material constants;
(b) A sensor which acquires the EM-wave information from the surroundings;
(c) A chip-type microcomputer to control the optimal condition for a material constant.

The ACMM is composed of producing and integrating these functions on the same board. By the way, devices which can electrically control the EM-wave beam, reflection, and transmission and absorption characteristics have been proposed so far. For example, Lee and Fong's paper, which discussed the EM-wave scattering using a negative resistance diode, was said to be a pioneering research in these materials of active type [8]. Further, in addition to

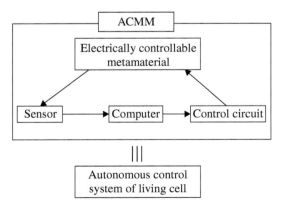

Figure 10.1 Fundamental concept of ACMM construction.

the beam control antenna devices [9, 10], phase shifters [11], phase modulation planar screens [12], and EM-wave devices employing the idea of active circuit elements were also proposed.

Also, the application of active circuit elements has been extensively studied, especially in the field of application to frequency selective surface (FSS) [13–15]. Among these papers, we can often find that FSSs are also called EM-wave absorbers [15]. Electrically controllable absorbers for radar applications have also been proposed [16].

Hereafter, a new electrically controllable EM-wave absorber, which can satisfy all the EM-wave absorption characteristics required for the EM-wave absorber, which is different from the above-described conventional configuration, is described in detail.

10.2 Configurations of the ACMM Absorber

This section describes the configuration method in the absorber part and discusses how to apply some of the active element circuits to use as the electromagnetic-wave absorber [17]. This absorber function part of the ACMM is referred to as an ACMM-type absorber.

In this absorber, diodes as the active elements are loaded in a central portion of the line-shaped rectangular printed circuits, as shown in Figure 10.2. The part of a square or rectangular shape, where the active elements are mounted, is called a unit cell. Using a substrate composed of unit cells which are arranged periodically, the basic ACMM-type absorber can be constructed by placing a spacer between the substrate and the conductive plates, as shown in Figure 10.3. Of course, when a conductive plate is not attached to the back of the spacer, this configuration can be applied to the spatial filter in the same way as the FSS. Figure 10.3a is a two-dimensional unit cell, and Figure 10.3b is a configuration example of a three-dimensional unit cell.

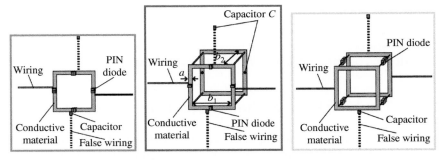

Figure 10.2 Constructions of two-dimensional and three-dimensional unit cells.

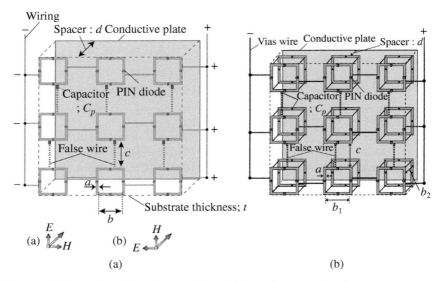

Figure 10.3 Examples of absorber consisting of (a) two-dimensional and (b) three-dimensional unit cells.

Table 10.1 Dimension example of a two-dimensional unit cell.

a (mm)	b (mm)	c (mm)	t (mm)
2.0	20.0	25.0	0.8

In these configurations, the factors controlling the equivalent material constants depend on the following:

(a) Shape effect of the unit cells,
(b) Arrangement form of the unit cell corresponding to the Bravais lattice arrangement,
(c) Circuit characteristics of the unit cell itself.
   This type of EM-wave absorber behaves as if the equivalent specific permittivity and permeability are newly given due to these factors.

10.3 The Main Point as the Technical Breakthrough

As shown in Figure 10.4, the problem of complex wiring to the diodes has been solved by introducing the new concept of "false wiring." As a result, initially, the number of power supply wiring per square substrate of 30 cm × 30 cm was 144, but by introducing the idea of false wiring, this problem could be solved

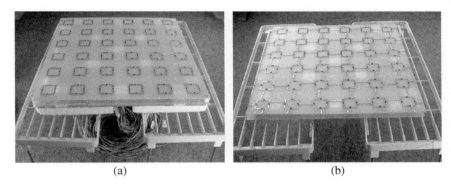

(a)   (b)

Figure 10.4 (a) Photograph of the initial substrate with 144 feeder lines on one board. (b) Photograph of improved substrate using only two feeder lines.

using only two wires from both sides of the substrate. By breaking through this feeder wiring problem, the present absorber implementation becomes possible in actual application fields of the absorbers.

By the way, in order to design this EM-wave absorber at a certain frequency, the method of using an finite difference time domain (FDTD) analysis to determine the design dimensions of each part was conducted as to the 3D unit cell analysis without using computer simulation. Figure 10.5 exhibits the basic structure and dimensions of each part using a three-dimensional unit cell when targeting a wireless LAN frequency (5.2 GHz).

The relative permittivity of the substrates used here is all 3.0 approximately. The main feeder wirings for the PIN diode are arranged on the right and left of the substrate. Vertical bias wirings show the so-called false wirings (or pseudo wirings) and are connected between conductive square circuits.

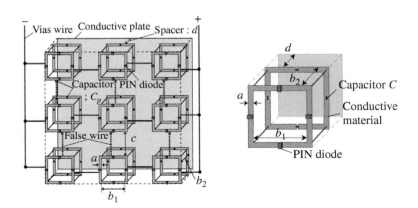

Figure 10.5 Architecture of the 3D unit cell microwave absorber designed using FDTD analysis.

Table 10.2 Dimensions of a three-dimensional unit cell.

a (mm)	2.0	b_1 (mm)	20.0
b_2 (mm)	12.0	c (mm)	25.0
d (mm)	5.0	C (pF)	1.0

10.4   Characteristics as the EM-Wave Absorber

This section describes that the characteristics of the EM-wave absorber, composed of two- and three-dimensional unit cells, are based on the above-mentioned idea, while pointing out each investigation point. Since this type of absorber has a shape of construction different from the conventional ones, several problems have to be solved in practical use, as is shown in Figure 10.6 [17].

These are summarized as follows:

(a) The complicated wiring problems,
(b) The problem of the EM-wave absorber characteristic control by controlling the bias voltage in the active element,

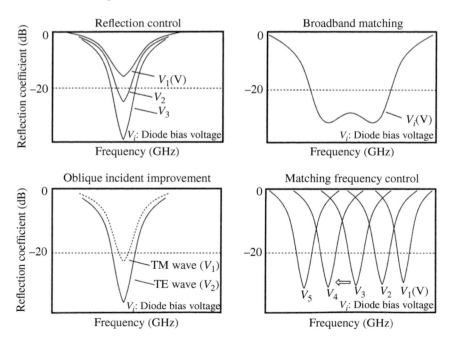

Figure 10.6 Characteristics to be satisfied by the absorbers.

(c) Stability of the absorption characteristics of waves from the incident EM-wave field from any direction,

(d) Oblique incidence characteristic problems,

(e) Control of the EM-wave absorption frequency,

(f) Wideband absorbing characteristic, and so on.

10.4.1 Complicated Wiring Problems

With respect to the task of (a) wiring, it is solved by the idea of pseudo wirings, as shown in Figure 10.3, and the breakthrough for practical use is achieved.

10.4.2 Controlling the Problem of Absorber Characteristics

Figure 10.7a shows the case where the incident electric-field direction is parallel to the pseudo wiring and (b) shows the case where the incident

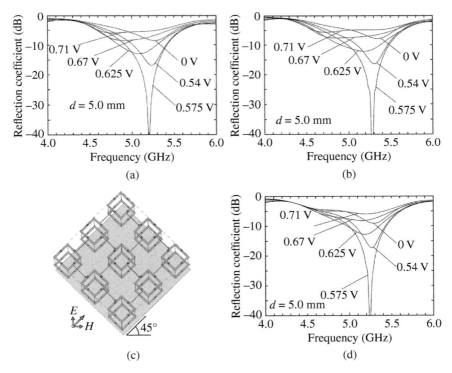

Figure 10.7 Measurement results of the incidence characteristics of the three-dimensional unit cell ACMM absorber. (a) Absorbing characteristic when electric field direction is parallel to the pseudo wiring. (b) Absorbing characteristic when electric field direction is vertical to the pseudo wiring. (c) Figure meaning that the electric field is incident on the substrate being rotated 45°. (d) Absorption characteristics in the case of (c).

electric-field direction is in the direction vertical to the pseudo wiring. In this way, by controlling the voltage of the diode, excellent EM-wave absorption characteristics can be realized independent of the incident wave polarization characteristics, as shown in Figure 10.7. The bias voltage value in the figure means the applied voltage when converted into one diode.

10.4.3 Stability of Wave Absorption Characteristics

Next, regarding the problem in (c), the stability of the EM-wave absorption characteristics when incident waves come from arbitrary directions to the absorber should be examined. Figure 10.7d shows the EM-wave absorber characteristic in a case where a rectangular EM-wave absorber is rotated by 45°, as shown in Figure 10.7c. It can be seen from this data that the EM-wave absorption characteristic is stable to changes in its incidence direction, when the absorber characteristic is evaluated from the value of the reflection coefficient −20 dB.

10.4.4 Oblique Incidence Characteristics

As for the problem of oblique incidence characteristics, this has been improved by devising the structure forms of a unit cell [17]. In this type of EM-wave absorber, it should be noted that the EM-wave absorption characteristics especially deteriorate for oblique incidence, and the EM-wave absorption

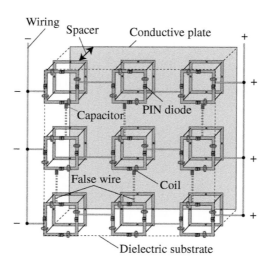

Figure 10.8 A new configuration of ACMM absorber composed of 3D unit cells with installed PIN diodes, capacitors, and inductors.

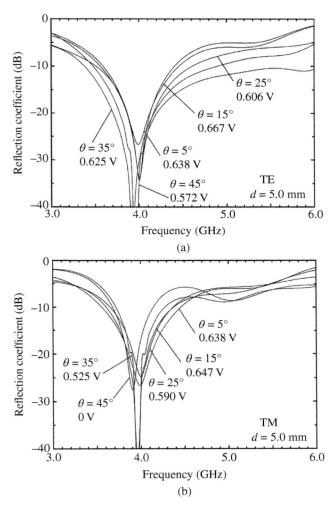

Figure 10.9 Absorption characteristics of the (a) TE and (b) TM waves at oblique incidence [17].

frequency may cause different problems between the TE and the TM waves [17].

As an example of a solution to this problem, as shown in Figure 10.8, small coils are mounted on unit circuits. Figure 10.9 shows the oblique incident characteristics of both the TE and TM waves when applying these countermeasures [17]. As long as it is evaluated with a reflection coefficient of −20 dB, good oblique incidence characteristics are attained.

10.4.5 Controllability of Frequency Characteristic

Regarding problem (e), in combination with the double-layer board configuration, as shown in Figure 10.10, it becomes possible to continuously change the matching frequency characteristics [7]. As an example, Figure 10.10a shows the case where each unit cell is configured in two and three dimensions. The first layer consists of two-dimensional unit cells with PIN diodes in the middle of square conductive circuits, as shown in Figure 10.10a. The second layer consists of three-dimensional unit cells with PIN diodes in the direction perpendicular to a substrate surface, as shown in Figure 10.10a,b respectively.

The sizes of each square unit cell in this case are the same as in Tables 10.1 and 10.2. The bias voltages V_1 and V_2 are supplied to each layer of the PIN diodes. That is, the bias voltage of each layer is independent. Figure 10.11 shows the matching characteristic for the TE wave in the present double-layered ACMM absorber. V_1 expresses the bias voltages of the diode on the first layer and V_2 on the second layer. By gradually applying the bias voltage to the diodes, the EM-wave absorption characteristic can be shifted toward the lower frequency region, while maintaining the matching characteristic −20 dB or less, as shown in Figure 10.11. In this way, it is found that the matching frequency can be changed over a wide bandwidth broader than over 1.3 GHz by controlling the PIN diode bias voltages on both the first and second layers.

While this configuration is fairly complex, one advantage of this type of material is that it can extend the same design concept from the low-frequency region to the high-frequency region by means such as wavelength shortening, for example.

10.4.6 Broadband Characteristic

The configuration of this absorber is shown in Figure 10.12 [17]. This EM-wave absorber has a two-layer plate structure composed of two-dimensional unit cells. When looking at Figure 10.12 from the front, the circuit pattern of the first layer has the same configuration dimensions as the two-dimensional ACMM-type absorber described earlier. In the substrate surface layer, rectangular conductor patches shown in black are alternately loaded in order to obtain broadband characteristics. The first layer is composed of a one-dimensional unit cell. The length of one side of the square conductive patch is 17.5 mm. The second layer consists of a substrate of a passive circuit element having a microchip resistor with 150 Ω.

The dimension of the circuit pattern of the second layer is the same as its dimension on the first surface substrate. The spacing d_1 and d_2 between both boards are 12 and 5 mm, respectively. An example of this wave absorption characteristic as to the case of dimensions in Table 10.3 is shown in Figure 10.13 [17].

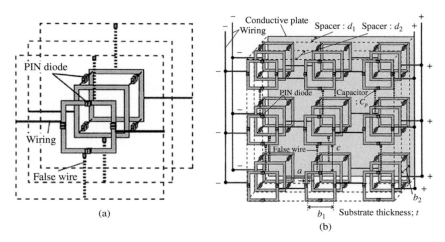

(a)

(b)

Figure 10.10 Configuration of the absorber for controlling the EM-wave absorption frequency.

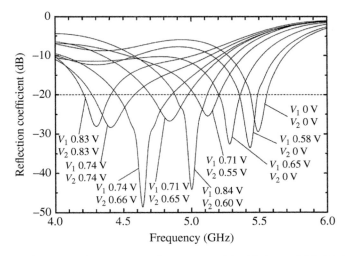

Figure 10.11 Absorption characteristics of a new type of double-layered ACMM absorber (Frequency changing characteristics of TE-wave case [7]).

Figure 10.13a shows the absorbing characteristics in a normal incident wave, in which the electric field is parallel to the pseudo wirings. It is found that when the allowable reflection coefficient is assumed to be −20 dB, broadband absorption characteristics having an absorption bandwidth of 1.3 GHz can be achieved. Table 10.4 is a summary of the specifications when

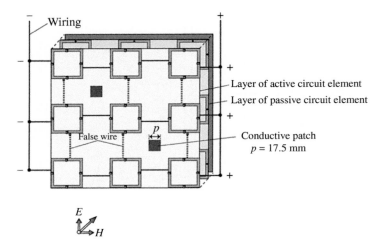

Figure 10.12 Configuration of a broad bandwidth absorber. (The dimension of the spacer is $d_1 = 12.0$ mm and $d_2 = 5.0$ mm, respectively. See Figure 10.10 for the intervals d_1 and d_2.)

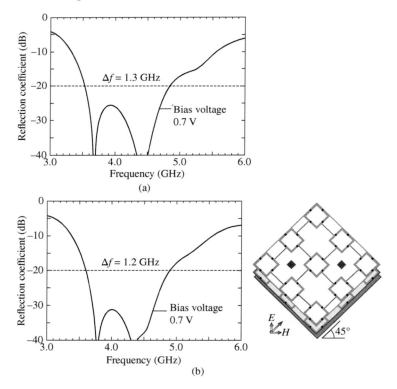

Figure 10.13 Example of the broadband absorbing characteristics of the ACMM-type absorber. (a) When the electric-field and the magnetic-field components are parallel to each side of the rectangular substrate. (b) When electric-field or magnetic-field component is incident on each side of a rectangular substrate at an angle of 45°.

Table 10.3 Configuration dimensions in a broadband absorber of ACMM type.

a (mm)	2.0	b (mm)	20.0
c (mm)	25.0	t (mm)	0.8
C_p (pF)	1.0	R (Ω)	150

Table 10.4 Specifications of ACMM-type absorber characteristics.

Absorber type	Bias voltage (V)	Total thickness (mm)	Matching frequency f_0 (GHz)	Bandwidth Δf (GHz)	−20 dB fractional bandwidth (%)
Two-dimensional absorber	0.59	17.4	4.82	0.2	4.1
Three-dimensional absorber	0.575	17.0	5.2	0.12	2.3
Two-layer-type absorber without patch	0.625	17.0	4.68	0.64	13.7
Two-layer-type absorber with patch	0.7	17.0	4.19	1.3	31

an ACMM-type EM-wave absorber consisting of a two-dimensional unit element or a three-dimensional unit element, and the wideband-type absorber in the present section is loaded with and without a rectangular patch.

10.5 Input Impedance Characteristic

In this section, let us briefly describe the actual results of measuring the input impedance in the Smith chart to understand the absorption mechanism of the ACMM-type absorber.

Figure 10.14 shows the input impedance characteristics of the ACMM-type absorber. This is the case of using a three-dimensional unit cell and the data obtained by measuring how the input impedance varies with the frequency change in the case of a vertical incidence. From these figures, it can be perceived that the electromagnetic-wave absorption characteristic passes to the absorbing state through the locus on the trochoid with respect to the change in the bias voltage of the diode. As described earlier, the present ACMM-type absorber has a great feature, since it can be configured to satisfy all the conditions that must

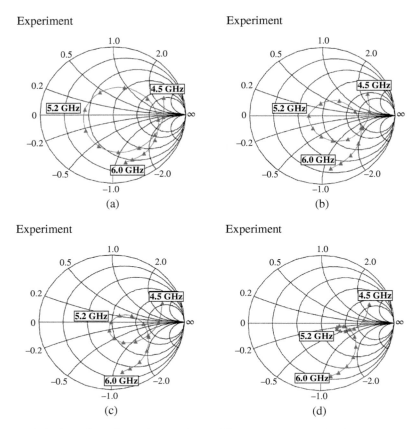

Figure 10.14 Input impedance characteristics of the ACMM-3D-type absorber. (a) Bias voltage 0 V, (b) bias voltage 0.567 V, (c) bias voltage 0.575 V, and (d) bias voltage 0.625 V.

be provided in an EM-wave absorber. In addition, this ACMM-type absorber can be configured with a relatively large unit cell in a frequently used microwave band, and this is also preferable from the viewpoint of easily designing the absorber.

Furthermore, the configuration of the present absorber in the high-frequency range is also made possible by the introduction of integrated circuit technology, while maintaining the same form based on the concept of wavelength shortening, as described in the previous section.

As described earlier, the present absorber can be designed to satisfy all the conditions that are required in an EM-wave absorber depending on application fields.

10.6   Examples of Application Fields

This ACMM-type absorber has a large variety of applications, not only as an EM-wave absorber but also in antenna applications, communication systems for use in moving vehicle control, radio wave signals (microwave signals), and the like. If this wave absorber is applied to a conventional anechoic chamber, a shield room, or a reflective box, it will be obvious that various facilities with multifaceted functions can be constructed.

Here, we briefly introduce the diversity of application fields utilizing this wave absorber.

In the present EM-wave absorber, we can turn on and off the diode voltage to adjust the operation of this EM-wave absorber for repeated absorption and reflection at constant intervals. Therefore, with this configuration, the unmodulated waves incident on the wave absorber are reradiated as a pulse-modulated wave having a constant period. That is, the unmodulated wave emitted from the transmitting side can be converted into a pulse-modulated wave with a predetermined frequency and can be received by the transmitting side. In this communication system, depending on a particular characteristic with the present wave absorber, it becomes possible to make a characteristic that a modulated wave can be received at the front of the wave absorber only for a vertical incident wave to an EM-wave absorber. Applying this fact, communication becomes possible only when the transmitted wave comes to the front of the EM-wave absorber; and as a result, it becomes possible to realize a localized positioning communication system.

As is well known, GPS plays an important role globally as a predetermined position communication system. On the other hand, the present communication system has the feature that it can be used for communication in a local field such as an underground shopping center or a long tunnel. In this sense, this communication method is called a Localized Positioning Communication System (LPCS).

A major feature of this communication system is that it is easy to install on mobile bodies, because the constituent device is simple and lightweight. Further, its various kinds of application fields are expected.

Normally, in general communication, antennas are employed for transmitting and receiving signals. However, the antenna always has side lobe characteristics, which makes it difficult to implement localized communication.

Hence, from the perspective of supplementing this kind of technology, the "raison d'être" of the Localized Positioning Communication System based on the present principle can be granted.

Let us here concretely describe the application examples based on the proposed idea.

Example 10.1   EM-Wave Signal System.

When taking into account the abnormal weather age, to spread an automatic control car and a drone, or a car with a flying function, etc., which are developed as future technologies, instead of the conventional signals that have to be visually observed, microwave signals will also be needed. The application of this principle of a localized communication system, which makes it possible to avoid problems with antenna side lobes, will enable the applications to some traffic EM-wave signals, being called YK signal from its principle.

Example 10.2   Application to Code Sensing Communication Systems

The author has proposed a communication method, called the code sensing communication system (COSCOS) [18, 19], in which EM waves having frequencies $f_1, f_2, f_3, \ldots$ are irradiated in the form of beams instead of bar codes, and this code information is read out while the mobile body moves. This can be easily understood when thinking that this is an application example of the principle of a localized positioning communication system.

Namely, by disposing the EM-wave absorber in the form of a bar code, radiating an unmodulated wave while the mobile body moves, and reading the signal of the frequency sequence of the modulated $f_1, f_2, f_3, \ldots$, we can receive the information of a local position. This communication method is considered to be effective for high-speed mobile communication and also has high confidentiality.

Example 10.3   Application to an Antenna to Control the Beam Width

If this EM-wave absorber is applied to antennas with various reflective surfaces, it is possible to control the beam width of the antenna and enable communication while taking the electromagnetic environment into consideration. Consequently, the application range of this wave absorber is extremely wide, and can be considered not only as the use of an EM-wave absorber but also a technology that can be assimilated with communication technologies that are increasingly employing a higher frequency and with AI technologies.

References

1 Kotsuka, Y. (organizer or chair) (1995). *Biological Effect of Electromagnetic Fields and Measurements*,", *IEEJ Measurement Committee*, 4. Corona Publishing Co. Ltd.

2 Kotsuka, Y. and Amano, M. (2003). Microwave functional material for EMC. In: *Technical Report of IEICE, EMCJ2003-40*, 13–18.

3 Kotsuka, Y. and Amano, M. (2004). A new EM-wave absorber using functional electromagnetic cell material. In: *Proceedings of EMC' 04 Sendai*, vol. 1, 301–304.

4 Kotsuka, Y. and Amano, M. (2004). A new concept for functional electromagnetic cell material for microwave and millimeter use. In: *2004 IEEE MTT-S International Microwave Symposium Digest*, 253–256.

5 Kotsuka, Y., Sugiyama, S., and Kawamura, C. (2006). New method of constructing computer controllable metamaterial and its microwave absorber application. In: *2006 IEEE MTT-S International Microwave Symposium Digest*, 927–930.

6 Kotsuka, Y., Sugiyama, S., Kawamura, C., and Murano, K. (2007). Novel computer control metamaterial beyond conventional configuration and its microwave absorber application. In: *2007 IEEE MTT-S International Microwave Symposium Digest*, 1627–1630.

7 Kotsuka, Y. and Kawamura, C. (2007). Novel metamaterial based on the concept of autonomous control system of living cell and its EMC applications. In: *2007 IEEE EMC-S International EMC Symposium Digest*, WE-PM2-SS4.

8 Lee, W. and Fong, T.T. (1972). Electromagnetic wave scattering from an active corrugated structure. *J. Appl. Phys.* 33: 388–396.

9 Alexopoulos, N.G., Uslenghi, P.L.E., and Tadler, G.A. (1974). Antenna beam scanning by active impedance loading. *IEEE Trans. Antennas Propag.* AP-22: 722–723.

10 Chekroun, C., Herrick, D., Michel, Y. et al. (1981). New method of electronic scanning. *Microwave J.* 45–53.

11 Lam, W.W., Jou, C.F., Chen, H.Z. et al. (1988). Millimeter-wave diode-grid phase shifters. *IEEE Trans. Microwave Theory Tech.* 36 (5): 902–907.

12 Tannant, A. and Chambers, B. (1998). Experimental phase modulating planar screen. *Electronics Lett.* 34 (11): 1143–1144.

13 Cahill, B.M. and Parker, E.A. (2001). Field switching in an enclosure with active FSS screen. *Electron. Lett.* 37 (4).

14 Chang, T.K. and Langley, R.J. (1993). An active square loop frequency selective surface. *IEEE Microwave Guided Wave Lett.* 3 (10): 387–388.

15 Chakravarty, S., Mitra, R., and Williams, N.R. (2002). Application of a micromagnetic algorithm (MGA) to the design of broad-band microwave absorbers using multiple frequency selective surface screens buried in dielectrics. *IEEE Trans. Antennas Propag.* 50 (3): 284–296.

16 Chambers, B. and Ford, K.L. (2000). Topology for tunable radar absorbers. *Electron. Lett.* 36 (15).

17 Kotsuka, Y., Murano, K., Amano, M., and Sugiyama, S. (2010). Novel right-handed metamaterial based on the concept of autonomous control system of living cells and its absorber applications. *IEEE Trans. Electromagn. Compat.* 3: 556–565.

18 Kotsuka, Y. and Takahashi, T. (1989). New proposal on navigation system by reading information code. In: *National Convention in IEICE, B-160*, 2–160.

19 Kotsuka, Y. and Tsuji, K. (1993). An investigation of code-sensing communication system by radiation of multiple frequency beams. In: *National Convention in IEICE, B-788*, 3–339.

Index

Electromagnetic Wave Absorbers: Detailed Theories and Applications, First Edition. Youji Kotsuka.
© 2019 John Wiley & Sons, Inc. Published 2019 by John Wiley & Sons, Inc.